GLOBAL WARMING

The science of climate change

FRANCES DRAKE
School of Geography, University of Leeds

A member of the Hodder Headline Group
LONDON
Co-published in the United States of America
by Oxford University Press Inc., New York

First published in Great Britain in 2000 by Arnold,
a member of the Hodder Headline Group,
338 Euston Road, London NW1 3BH

http://www.arnoldpublishers.com

Co-published in the United States of America by
Oxford University Press Inc.,
198 Madison Avenue, New York, NY10016

British Library Cataloguing in Publication Data
A catalogue record for this book is available from the British Library

Library of Congress Cataloging-in-Publication Data
A catalog record for this book is available from the Library of Congress

ISBN 0 340 65301 9 (hb)
ISBN 0 340 65302 7 (pb)

1 2 3 4 5 6 7 8 9 10

Production Editor: Rada Radojicic
Production Controller: Fiona Byrne
Cover Design: Terry Griffiths

Typeset in 10/12 pt Sabon by
J&L Composition Ltd, Filey, North Yorkshire
Printed and bound in Great Britain by MPG Books Ltd

What do you you think about this book? Or any other Arnold title?
Please send your comments to feedback.arnold@hodder.co.uk

Contents

Preface

This book is an intermediate text book for students approaching their final years at university. The book concentrates on the climatological aspects of global warming and how the science is being used to inform policy-makers. It provides, I hope, a readable account of the scientific story so far. The aim is to introduce as many of the basic concepts behind global warming as possible so that the reader can then go on and explore the latest published scientific developments. This book is meant as a primer, for want of a better word, to this wider literature. Climate and meteorology have increasingly become the domain of mathematicians and physicists. Many of the research papers written in this field contain mathematical and physical concepts way beyond first year undergraduate level. This book is designed for geography, earth science and environmental science students and keeps the maths to a minimum. Greater detail can be found in the referenced textbooks and in the Further Reading list. The key scientific literature is also extensively referenced. I have also referenced less technical reviews in readily available journals like *New Scientist*. Websites are not generally referenced because they become outdated rapidly. A short list of useful Websites appears in an appendix. Keywords are shown in bold and defined in the Glossary.

The first two chapters provide an introduction to many of the physical concepts used. The book then goes on to place global warming within the wider context of climate change. This is important because it helps to explain why scientists are unable to say for certain whether global warming is happening or not. To understand global warming we cannot ignore natural variations in the climate; they are of vital importance and are discussed in detail. The book presents the controversy over the evidence for global warming, from both observed data and climate models. It highlights the sceptics' complaints about this evidence and gives the believers' responses. These chapters are meant to give the reader an understanding of the arguments and the counter-arguments. Finally the book considers the wider issues of science and society. The likely impacts of global warming and the ethical questions these raise are reviewed. The role of science in answering those questions is also examined alongside the policy implications. The last chapter may challenge some readers' perception of science and its ability to influence policy decisions. However, the role of science, and particularly scientific uncertainty in wider social constructs, is critical to understanding why global warming remains a contentious issue.

Frances Drake
30 September 1999

Acknowledgements

I must thank various people for helping to make this book possible: David Appleyard from the Graphics Unit, School of Geography, University of Leeds for drawing the diagrams; Anna Matthews for contributions to the last chapter and for reading the book; David Yeomans for proof-reading and jovial comments; Laura McKelvie, Luciana O'Flaherty and Emma Heyworth-Dunn from Arnold for encouragement and answering numerous queries. Finally thanks to my family for their continued support.

The author and publishers would like to thank the following for permission to use copyright material in this book:

American Geophysical Union (Figures 3.1 and 3.4); Arnold (Figures 3.7 and 6.6); Cambridge University Press (Figures 3.6 and 4.3; Tables 2.1 and 2.2); European Communities (Figures 5.1 and 7.1); International Institute for Applied System Analysis (Figure 7.2); Intergovernmental Panel on Climate Change (Figures 4.3, 5.2, 5.4, 5.6, 6.4, 7.3 and 7.4; Tables 2.3, 5.1 and 6.1); John Wiley and Sons Limited (Figure 6.3); Kluwer Academic Publishers and A.B. Pittock (Figure 5.8); Kluwer Academic/Plenum Publishers and T.H. McGovern (Figures 4.5 and 4.6); Macmillan Magazines Limited, Crown Copyright and J.F.B. Mitchell (Figure 6.5); The McGraw-Hill Companies (Figure 2.5); Routledge (Figures 5.5 and 5.7); Sage (Figure 8.1); Scientific American Inc. (Figure 4.2); Tellus Editorial Office (Figure 3.2); Thomson Learning Global Rights Group (Figure 1.10); World Meteorological Office (Figure 1.12).

Every effort has been made to trace the copyright holders of material produced in this book. Any rights not acknowledged here, and/or inappropriate attributions, will be rectified in subsequent editions if notice is given to the publisher.

Acronyms, abbreviations and chemical symbols

Where appropriate, an entry can be found in the Glossary under the full title.

AGCM: atmospheric general circulation model
AMIP: Atmospheric Model Intercomparison Project
AOGCM: coupled atmospheric and oceanic general circulation model also CGCM.
AoSIS: Association of Small Island States
BP: years before present
CBA: cost benefit analysis
CFCs: chlorofluorocarbons
CH_3Cl: methyl chloride
CH_4: methane
CO: carbon monoxide
CO_2: carbon dioxide
CoP: Conference of the Parties
DVI: dust veil index
EBM: energy balance model
EMT: Ecological Modernization Theory
ENSO: El Niño Southern Oscillation
EPA: Environmental Protection Agency
GCM: General circulation or general climate model
GISP: Greenland Ice Sheet Projects
GtC/yr: gigatonnes of carbon per year
GWP: global warming potential
HCFCs: hydrochlorofluorocarbons
HFCs: hydrofluorocarbons
IAM: integrated assessment model
IPCC: Intergovernmental Panel on Climate Change
ITCZ: Intertropical Convergence Zone
JUSCANZ: Japan, USA, Canada, Australia and New Zealand
MCA: Multicriteria analysis
NADW: North Atlantic deep water
NAO: North Atlantic Oscillation
NMHC: non-methane hydrocarbons

NMVOC: non-methane volatile organic compounds
NO: nitrogen oxide
NO$_2$: nitrogen dioxide
N$_2$O: nitrous oxide
O$_3$: ozone
ODP: ozone depletion potential
OECD: The Organization for Economic Cooperation and Development
OGCM: oceanic general circulation model
PFCs: perfluorocarbons
ppbv: parts per billion volume
ppmv: parts per million volume
pptv: parts per trillion volume
psu: practical salinity units
RCM: regional climate model
RST: Risk Society Theory
SF$_6$: sulphur hexafluoride
SO$_x$: sulphates
SOI: Southern Oscillation Index
SPCZ: South Pacific Convergence Zone
SST: sea surface temperature
UNEP: United Nations Environment Programme
UNFCCC: United Nations Framework Convention on Climate Change
WMO: World Meteorological Organization
WTP: willing (or willingness) to pay

Introduction

Global **warming** is a term widely used to describe a potentially dramatic rise in the annual average global surface temperature of the Earth. Estimates of how big that temperature increase will be range from $1.5\,^{\circ}C$ to $4\,^{\circ}C$ (Houghton *et al.*, 1996). If such a change in temperature occurs it is most likely that there will be other alterations to our climate. The rainfall distribution might change and the frequency of severe **weather** events, such as hurricanes and typhoons, might increase. All the variables that we use to describe the weather and **climate** might undergo a profound alteration. Most scientists believe that there is good evidence that this change is due to increasing amounts of carbon dioxide (CO_2) and other gases in the Earth's atmosphere. These greenhouse gases act to trap outgoing thermal radiation which then warms the Earth. But why is carbon dioxide increasing; what is the real cause? To answer that question we need only to look around at our homes, our schools and our workplaces. Since fire was discovered we have been burning fuels that release carbon dioxide into the atmosphere. With the onset of the industrial revolution in the 1700s, increasing use has been made of **fossil fuels** which release large amounts of carbon dioxide when burnt (Weyant and Yanigisawa, 1998). The industrial and domestic energy demands of our modern society mean that approximately 7 gigatonnes (Gt) of carbon from carbon dioxide is released every year (Houghton *et al.*, 1996). Consequently, as our technological ability has increased so has our potential for altering the environment in which we live.

With such profound consequences, and the culprit for the changes known, it might be thought negligent of the global community if no attempt was made to address the situation. Consequently, there have been world-wide efforts to tackle global warming. By 1988 the global scientific community had been canvassed and the consensus opinion was that the predicted changes in climate were significant enough to warrant action to curb greenhouse gases. The Toronto Conference in Canada that year led to the setting up of the Intergovernmental Panel on Climate Change (IPCC) by the World

Meteorological Organization (WMO) and the United Nations Environment Programme (UNEP). The role of the IPCC was to examine the scientific evidence and report back to the world's leaders. There followed a series of high profile meetings aimed at addressing this problem which led to the Rio Earth Summit in 1992. At this meeting, the United Nations Framework Convention on Climate Change (UNFCCC) was adopted by the signatories. This was essentially a commitment to stabilize greenhouse gases in the atmosphere at a level that would not endanger the **climate system** and was interpreted by most signatories as meaning a return to 1990 emission levels of greenhouse gases by the year 2000 (Howes *et al.*, 1997). Most developed nations will fail to meet this reduction.

There were no targets set in 1992 and when the signatories met in Berlin in spring 1995 (First **Conference of the Parties** – CoP-1), little was agreed other than that emission targets for beyond the year 2000 needed to be set by 1997. In July 1996 the second Conference of Parties was held in Geneva (CoP-2) and this was more positive. The USA, perhaps the world's largest emitter of greenhouse gases, endorsed action (Howes *et al.*, 1997). By December 1997 and the third meeting in Japan (CoP-3), the Kyoto Protocol agreed targets for a group of six greenhouse gases (ENDS, 1997). An overall target, rather than individual targets for each gas, was set, thereby allowing rising emissions of one gas to be offset by extra reductions of another gas. It was agreed that developed countries would cut emissions by 5.2 per cent by 2008–2012. The USA agreed to a reduction of 7 per cent and the European Union (EU) to 8 per cent (ENDS, 1997). In November 1998, the fourth meeting in Buenos Aires (CoP-4) was to decide on the implementation strategy for the Kyoto Protocol. Instead, most of the decisions were left for a rolling programme over the next two years. Many issues were left unresolved and it is unclear whether enough has been done to stabilize the climate (Pearce, 1998). By mid-1999 the USA was still refusing to cut its own emissions until certain key developing countries accept greenhouse gas emission targets. Without some sort of resolution soon the whole policy process is in danger of stalling (Pearce, 1999). It looks increasingly likely that we will be unable to stop global warming.

Why has so little been achieved despite dire warnings, increased public environmental commitment and exhortations for businesses to 'go green'? (see for example Wheatley, 1993). Partly the answer lies in the complexity of the problem; there is no quick fix. Every facet of life in the modern developed world seems to require the consumption of fossil fuels, but the most serious impediment to change is the perception that for increased economic growth there must be continuously rising energy consumption, and consequently increasing greenhouse gas emissions. The argument is that to put the brakes on carbon dioxide emissions will seriously impede our economic welfare (Nordhaus and Yang, 1996). Some of the most powerful lobbying groups in the world have been highly successful at getting this message across, not only to politicians and policy-makers, but also to the public. There are many reasons for that success. One is our fear of economic recession and another is the skill with which the scientific uncertainty that surrounds the problem has been highlighted. The science has been interpreted to serve a range of vested interests and the disparity between the different messages presents a confused image of global warming to the public. Therefore it is important to understand the science of global warming and climate change so that we can appreciate the nature of the problem and the difficulties of providing scientific certainty. This then allows us to assess the claims and counter-claims of scientists, and also to appreciate the much wider arguments of economists, social scientists and politicians.

It has been said that global warming is the largest experiment ever undertaken by humankind (Revelle and Suess, 1957). It began unintentionally, with no idea of the end result and probably no way of stopping it. The experiment involves the whole planet and will affect not just our generation but many generations to come. The task is to understand that experiment, to try and predict what will happen and estimate just how big the resulting problems might be. An experiment that involves the whole Earth and everything on it is likely to be complex; we need to build our understanding on some solid foundations. The first two chapters will define what scientists mean by climate and discuss how it functions. Later chapters will review some of the evidence for and against the idea of global warming. As yet global warming is not certain and there are a few scientists who think that climate change is unlikely. At the end of the book some of the responses that society might take, given the scientific evidence, are discussed. Finally, the role of science and scientists within the global warming debate will be reviewed.

References

ENDS 1997: The unfinished climate business after Kyoto. *ENDS report no. 275*, 16–20.

Houghton, J.T., Meira Filho, L.G., Callander, B.A., Harris, N., Kattenberg, A. and Maskell, K. (eds) 1996: *Climate change 1995: the science of climate change*. Cambridge: Cambridge University Press.

Howes, R., Skea, J. and Whelan, B. 1997: *Clean and competitive: motivating environmental performance in industry*. London: Earthscan.

Nordhaus, W.D. and Yang, Z. 1996: A regional dynamic general equilibrium model of alternative climate-change strategies. *The American Economic Review* 86, 741–65.

Pearce, F. 1998: Green futures. *New Scientist* 160 (no. 2161), 16.

Pearce, F. 1999: A carbon fix? *New Scientist* 162 (no. 2190), 22.

Revelle, R. and Suess, H.E. 1957: Carbon dioxide exchange between atmosphere and ocean and the question of an increase of atmospheric CO_2 during the past decades. Tellus 9, 18–27.

Weyant, J. and Yanigisawa, Y. 1998: Energy and industry. In Raynor, S. and Malone, E.L. (eds), *Human choice and climate change: volume 2 resources and technology*. Columbus: Batelle Press.

Wheatley, M. 1993: *Green business: making it work for you*. London: Pitman Publishing.

|1|

Basic physical concepts

Climate and weather

The Earth continually receives energy from the Sun. If there was not something to balance this incoming energy then the Earth would always be heating up. The Earth, however, emits energy out into space, often called terrestrial energy, so it loses heat. One might expect that the outgoing energy at each location on the globe would have to exactly match that coming in, in order to maintain a balance, however, observations show that this is not the case. The latitudinal distribution of solar energy reveals that it is highest at the equator and then declines to a minimum at the poles. On an annual average the equator receives around 2.5 times as much energy from the Sun as the poles. Terrestrial radiation on the other hand is much more evenly distributed (see Fig. 1.1). At low latitudes, between 35° S and 40° N there is an excess of incoming energy and polewards of these latitudes there is a deficit. If there was nothing else, the low latitudes would be continually heating up and the polar regions would be cooling. This is not the case and there is no overall heating or cooling. Therefore the energy lost is balanced by energy gained. This is called thermal equilibrium. The Earth achieves this thermal equilibrium through the atmosphere and oceans. In response to the energy imbalance, atmospheric and oceanic circulations occur which redistribute the energy from the equator to the poles. It is the measurement of these circulations that define and quantify the weather and climate of a location. However, many elements can modify these large-scale circulations: the topography, the vegetation, and the land surface; these too require consideration.

The climate and weather are the result of complex interactions between components which together form (and define) the climate system. These components are the **atmosphere**, **hydrosphere** (free water, i.e. liquid or vapour), **cryosphere** (frozen water), **biosphere** and **lithosphere**. Weather is usually defined as the condition of the atmosphere at a particular location at a particular time. Climate is sometimes loosely defined as the average weather

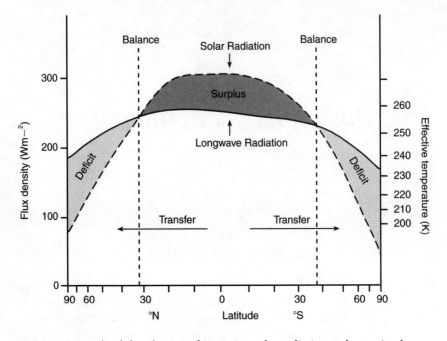

Figure 1.1 Latitudinal distribution of incoming solar radiation and outgoing long-wave radiation over the Earth's surface

conditions of a place. This rather loose definition, however, is not good enough for reasons that will be considered later on in this book. The climate is best defined as the average state of the various components of the climate system, together with the variability of those components, the averages being taken over periods ranging from years to centuries (Houghton *et al.*, 1990). An important point to recognize is that this definition of climate refers to its variability; climate is not static. Also as scientists learn more about the weather it has been realized that in some circumstances just considering the atmosphere is not enough. To accurately forecast the weather the oceans have to be considered as well.

The climate system

A system is a set of parts or components that are linked in an organized way. The components operate as a complex whole and result in some observed behaviour. By being part of the system the components are affected by it. Also, if a component is removed from a system the system is changed in some way. The components of the climate system are the atmosphere, hydrosphere, cryosphere, biosphere and lithosphere. Each component has a variety of properties usually referred to as variables. For example, some of the variables associated with the atmospheric component are temperature and pressure.

Usually it is the value the various variables take at some particular time that is of concern; this is called the system state. The number of variables to be considered is determined by the length of time chosen to study the system. Over short time-scales some variables may hardly alter and so they can effectively be considered as constant. For longer time-scales, however, more variables will have to be studied as more will change. This is particularly true of the climate system.

As a system changes from one state to another state it is said to be in transition. Some systems jump from one to state to another and can only exist in one state at any one time. This is called a discrete system. A good example of a discrete system is channel hopping on a TV remote control. Other systems move gradually from one system state to another and these are called continuous systems. The climate system is an example of a continuous system. The links which bind the climate system together are the transfers of energy and mass. What makes up a system is rather arbitrary. A system is usually a collection of objects that human beings want to study. Systems can be classified in terms of how they relate to their outside environment (Lockwood, 1979).

1 Isolated systems: These systems do not transfer energy or mass to, or from, their surrounding environment.
2 Closed system: Exchanges energy with its surrounding environment but there is no exchange of matter.
3 Open systems: These exchange both mass and energy with their surrounding environment.

Examples of isolated systems in the real world are rare. They are much more likely to be located in a laboratory. The usual example is of a closed test tube containing a gas. The mass and energy remain locked within the test tube. If there are any temperature gradients within the gas these will eventually even out. A very good example of a closed system is the planet Earth in space. There is some loss of mass through the atmosphere escaping into space but it is so small as to be negligible. There is, however, energy exchange, the Earth receives energy from the Sun and also loses energy to space.

Open systems are very common in the natural environment. River catchments, hurricanes and the climate system are all examples of such systems (Lockwood, 1979). The open system can be in dynamic equilibrium. This means that over long enough time-scales the features of the system appear constant but at any moment may vary from that. Open systems can be divided into three categories: decaying, haphazardly fluctuating and cascading. While some open systems will always belong to one category, some will change categories. Decaying systems use up the energy, and/or mass that are feeding them. As the name suggests, haphazardly fluctuating systems change in an unpredictable and random way. A cascading system is one which is composed of a series of subsystems. Mass and energy cascade from one subsystem to the next, so the output from one forms the input for the next.

The climate system is such a cascading system, the subsystems being the individual components of the climate system.

Energy

Energy is very important to the climate system. It is energy which drives the system and is constantly being exchanged from one component to the next. Energy is defined as the ability to do work and in physics work occurs when an object moves over some distance due to a force acting along the line of the movement. Sears, Zemansky and Young (1978) put forward the following example of work. Supposing someone asked you to move a heavy box for them. If you were to push it along a level floor you would be exerting a force on the box and causing it to move. Some component of the force would be directed in the direction of motion; this is work as defined in terms of physics. However, if you picked up the box and carried it you might well think that you were working hard but in physics terms work would only have been done when you picked up and put down the box. While carrying the box no work will have been done as the supporting force (supplied by you!) is in the vertical while the box is moving horizontally. In this case there is no component of the force acting in the direction of movement.

Energy comes in lots of different forms and it can transform from one type to another. A system has an energy budget and because energy can neither be created nor destroyed, the amount of energy gained by the system must equal the amount of energy lost plus any change in the stored energy of the system.

Potential energy is the amount of energy stored by an object. There are various different forms of potential energy. Gravitational potential energy is the energy an object possesses due to its position. It depends on mass, gravity and height. Essentially the higher and heavier an object is, the more potential energy it has. Chemical potential energy is the work that can be done due to a chemical change within a substance. The energy we get from food is stored as chemical potential energy. Similarly, fossil fuels have chemical potential energy which is transformed to provide heat.

Kinetic energy is the energy that an object possesses due to its movement. It depends on both the mass of the object and its speed. Objects moving at a higher speed have a higher kinetic energy than those moving at slower speeds. Those objects which have more mass have a greater kinetic energy than those with less mass. In a cup of water, where some molecules will be moving faster than others, there will be a range of speeds and hence a range of kinetic energies.

Temperature is defined as the mean kinetic energy per molecule of all the molecules in an object: in essence the average speed of the molecules of a substance. Temperature can be measured on a variety of different scales and the two used internationally are Celsius (also called centigrade) denoted by °C and Kelvin denoted by a K. The intervals on these two scales are the same, that is a temperature rise of 1°C is equivalent to a rise of 1 K. The definition of the zero points, however, are different on each scale. On the Celsius scale 0°C is the freezing point of water whereas on the Kelvin scale 0 K (around −273.15°C) is defined as the temperature at which all molecular motion stops. Bohren (1987) states that it is more true to say that matter ceases to radiate, rather than motion ceases. Only in an **ideal gas** would motion stop. An ideal gas is, a hypothetical one, where the molecules are far enough apart so as not to interact with each other. The atmosphere approximates to an ideal gas at room temperature. At **absolute zero** (0 K), quantum, rather than classical mechanics takes over. Any object above 0 K will have kinetic energy. Internal energy is the amount of kinetic and potential energy the molecules of an object contain.

Energy can be transferred by the means of work. This would occur if a mechanical force was used to compress a gas. Once compressed the gas can do more work so energy has been transferred to it through work. An example of this is a bicycle pump. You provide a mechanical force which compresses the air in the pump, causing it to become hotter, increasing its amount of kinetic energy. Another way in which energy is transferred from one object to another is by heat due to a temperature difference. The result of this energy transfer will be changes in the properties of the object. The temperature may rise, or the object may change state (as energy is being added the state change will be to a higher one from solid to liquid or liquid to gas) or both. A temperature rise is referred to as sensible heat exchange, whereas a change of state is known as **latent** (hidden) **heat**.

Specific heat capacity is the heat required to raise the temperature of 1g mass of a substance by 1°C. Different substances require different amounts of heat to raise their temperature by 1°C. The term calorie (cal) is defined as the heat needed to raise 1g of water from 14.5°C to 15.5°C. The specific heat capacity of water is taken to be 1 cal g^{-1} (°C)$^{-1}$. Soils have a much lower specific heat capacity than water (about 0.2 cal). Consequently although water takes longer to heat up than the land, it also takes longer to cool down. This property has an effect on the world's climate. The interior of continents such as Europe and Asia experience much more extreme summers and winters than coastal regions. At the coast the oceans act as a thermal store, smoothing out the difference between winter and summer.

Latent heat is the heat required for 1 g of mass of substance to change from one state to another without a rise in temperature. Using water as an example, ice is the solid state. To melt ice into a liquid requires energy as the water molecules need to move more quickly. Turning the liquid water into a vapour also requires energy. So in both these cases energy is absorbed. Perspiration is the body's way of using latent heat to cool us down. The layer of perspiration on a hot body uses that energy to turn from a liquid to a vapour taking away energy, so cooling us down. When a gas becomes a liquid or a liquid becomes a solid there is a consequent release of energy.

The transference of energy by heat can be by any of three processes:

conduction, convection and radiation. Conduction is the transfer of energy by molecular collisions. For example, if you take a long metal rod and heat it at one end, the energy is conducted to the other end of the rod. The molecules at the heated end of the rod will be moving fast due to the high temperature. They will collide into their cooler and more slowly moving neighbours and pass on some of their energy. This processes continues along the length of the rod to the unheated end. The molecules do not change their position along the rod. In contrast, convection involves the transferral of energy by the actual movement of the heated material. This commonly occurs in the lower atmosphere (troposphere) with heated parcels of air moving to colder parts of the atmosphere. Forced convection occurs when the material is forced to move, a common example in the atmosphere is the movement of air parcels up mountains. Free convection occurs when the material is at a different density to its surroundings. This is because the less dense an object, the lighter it is.

Latent heat release and convection play an important combined role in the atmosphere, transferring energy from low to high altitudes. To understand this consider what happens in the atmosphere by imagining a balloon filled with warm moist air slightly warmer than the surrounding atmosphere. The balloon is special, its thin surface skin effectively isolates the air from the surrounding atmosphere and stops any heat exchange with it. The atmospheric mass decreases with height so as the balloon rises there is less mass of atmosphere on it; less to compress it. The air inside the balloon will expand, doing work, so the air will cool. Eventually the air will cool sufficiently that the water vapour contained in it will condense out releasing latent heat. The air will continue to cool as it rises but the rate of cooling will not be as large, as it is now gaining energy from the latent heat release. This increases its ability to rise still higher. The atmosphere surrounding the balloon will also be cooling with height but so long as the air inside the balloon remains warmer than the surrounding atmosphere (i.e. cools at a slower rate), it will remain less dense than the atmosphere and will continue to rise. However, should the air cool faster than the surrounding atmosphere eventually it will become more dense than the atmosphere and begin to sink. As the balloon sinks the pressure will increase and the balloon will contract. The air inside will warm and any liquid water will begin to evaporate. As the air inside the balloon exchanged no heat with its surroundings the process is said to be **adiabatic**. The rate of change of temperature with height is known as the lapse rate and effectively the **lapse rate** of the atmosphere and that of the balloon or parcel of air will determine whether convective motions take place. So long as the parcel of air cools more slowly than the surrounding atmosphere (i.e. has a smaller lapse rate), then it will be less dense and more buoyant and convective motion will occur. This type of convective cell is particularly important in the tropics.

Radiation

The warmth you feel from the sun is due to radiation, similarly, the heat used in microwave cooking is transferred through radiation: microwaves. Radiation is propagated by electromagnetic waves. As the name suggests, these waves have both electrical and magnetic properties. They do not require a substance to travel through, unlike sound waves and in a vacuum they travel at a constant speed of $300\,000\,000$ ms^{-1} (3×108 ms^{-1}) called the speed of light.

Figure 1.2 illustrates some wave characteristics. The distance between the crest of each wave is called a wavelength, denoted by the Greek letter lambda (λ) and the height between a crest and a ridge is called the amplitude and is measured in meters. Frequency is the number of complete cycles per unit time, that is from a to b to a* and is denoted by the Greek letter nu (υ). Electromagnetic waves appear at all wavelengths, from very short X-rays to the long wavelengths of radio and television. This is called the electromagnetic spectrum and is shown in Figure 1.3. The typical unit of measurement for electromagnetic waves is the micron, a one millionth of a metre (μm).

The amount of radiation emitted by a unit surface area per unit time will depend on the type of surface and on its temperature. Any object with a temperature above absolute zero emits radiant energy. The radiant energy emitted occurs over a range of wavelengths called a spectrum. At low temperatures the amount of radiant energy emitted is small and occurs mostly at long wavelengths. As the temperature increases the total amount of energy radiated is increased and the radiation is at predominantly shorter wavelengths. At this point we might consider that if an object is continually emitting radiant energy surely it is transferring its own internal energy to its surroundings and will eventually cool down to absolute zero. That would be the case if energy were not supplied to the object. In an electric cooker hob the element glows and becomes hot due to energy being supplied to it electrically. Once that energy supply is switched off, it cools down rapidly to room temperature. It doesn't cool any further because the room itself is emitting energy which the cooking element can absorb. So as well as emitting energy all objects can also absorb energy. Absorption is a very important process and will be discussed in detail later. First, the concept of an ideal radiator will be introduced. In physics this is called a **black body**. A black body emits radiation such that at a temperature T it emits the maximum amount of radiation at all wavelengths. It also absorbs all the radiation that falls upon it at a temperature T. A good emitter is also a good absorber. The rate (or intensity) at which energy is emitted, F, can be calculated using the **Stefan-Boltzmann law:**

$$F = \sigma T^4 \tag{1.1}$$

The Greek letter sigma (σ) is the Stefan-Boltzmann constant which has a numerical value of 5.670×10^{-8} Wm^{-2}K^{-4} and T is the temperature of the surface in Kelvin. The Stefan-Boltzmann law shows that the rate of radiation

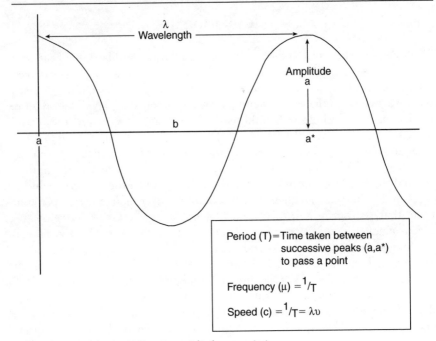

Figure 1.2 Simple wave diagram with characteristics

emission is proportional to the fourth power of the temperature of the object. So the higher the temperature, the more radiation is emitted per second. Although neither the Sun nor the Earth are black bodies, it is useful to look at the spectra that are produced by black bodies with the same temperature as the Sun and the Earth. These black body curves are shown in Figure 1.4. The temperature of the Sun is about 6000 K, therefore the energy output of the Sun is very energetic and occurs at short wavelengths. Another law called Wien's Law allows the calculation of the wavelength at which maximum radiation emission occurs, essentially the peak of the black body spectrum. Wien's Law states that

$$\lambda_{max} = b/T \qquad\qquad\qquad (1.2)$$

Where the temperature (T) is measured in Kelvin and b is a constant with a value of 2.898×10^{-3} mK. The maximum wavelength for the Sun is 0.5 μm. The spectrum of the Sun very closely resembles that of a black body and so the Stefan-Boltzmann law holds well. The spectrum generated by the Earth's surface at around 300 K is at lower wavelengths, thermal or infra-red wavelengths, typically the region of maximum radiation emission occurs at wavelengths around 10 μm. Thus there is very little overlap between the Sun's spectrum and the Earth's. This allows these two radiation streams to be treated separately, often referred to as the solar and terrestrial radiation streams, respectively. The Earth's spectrum does not approximate closely to a

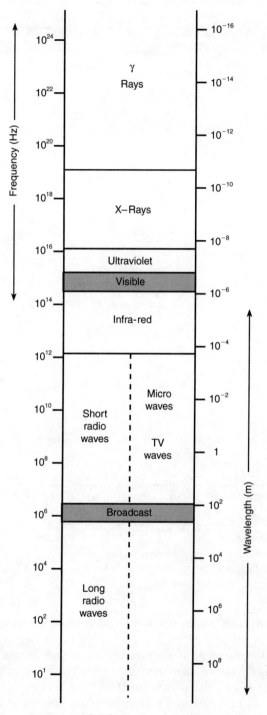

Figure 1.3 The electromagnetic spectrum

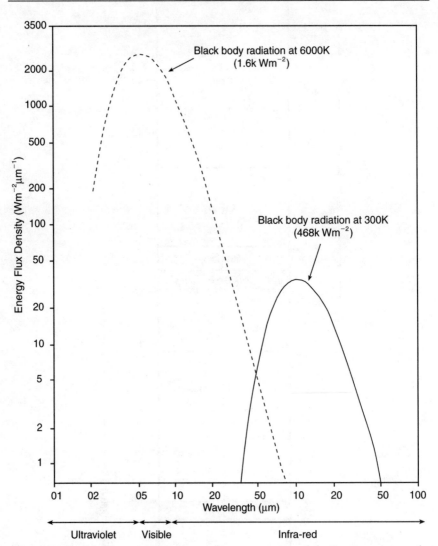

Figure 1.4 Black body curves for the emitting temperature of the Sun (approximately 6000 K) and for the Earth (approximately 300 K)

black body. Kirchoff's law defines how a real body compares to a black body:

$$\varepsilon = W/Wb \qquad\qquad (1.3)$$

The Greek letter epsilon (ε) denotes emissivity, W is the real emittance and Wb black body emittance. **Emissivity** is essentially a measure of how good an object is at emitting radiation and can take a value between zero and one. The emissivity of a black body must therefore be one. A body emitting no radiation (white body) is zero. All real radiators fall between these and are referred to as grey bodies. Usually the emission curves for grey bodies are far more complicated than the corresponding black body curve. Emissivity varies with the temperature and the nature of the surface and allows for the greyness of bodies. The emissivity of the Earth is about 0.7. A common refinement is made to the Stefan-Boltzmann Law to make an allowance for the fact that not all bodies act as perfect radiators.

$$F = \varepsilon \sigma T^4 \tag{1.4}$$

Processes affecting electromagnetic radiation

Electromagnetic radiation is everywhere, coming from the Sun in the form of heat and light, and from all the objects around us. What happens to that electromagnetic radiation when it reaches an object? That will depend on the wavelength of the incident light, the angle at which it strikes the surface and the nature of the surface, to name some of the most important. There are two important processes that can occur called **absorption** and **scattering**. These will be discussed after a brief review of the processes of reflection and refraction.

Refraction and reflection

To illustrate the process of refraction, consider visible light moving in the atmosphere then passing through glass. The light that passes through the glass is said to be transmitted. As the glass is more dense than the atmosphere the visible light will be slowed by the glass. Refraction occurs when the light strikes the glass at an angle and the path of the light bends. If the light strikes at 90° to the surface (called the normal) of the glass, then no bending occurs. The amount of refraction is given by Snell's Law and depends on the refractive index of the media (n) and the angle of incidence (θ). Reflection takes place when light striking a surface at an angle bounces off at the same angle. Therefore, reflection causes a change in the direction of the incident radiation but there is no change in any other properties. Figure 1.5 shows the incident, reflected, refracted and transmitted beam. Note that the beams are all in the same plane of incidence. Direct solar radiation is reflected and radiation which has been scattered by the atmosphere is also reflected. The reflection of scattered radiation can be quite significant. The nature of the reflected energy depends strongly on the surface roughness of the reflecting surface in relation to the wavelength of the incident light. The direction of the incident beam is also important.

Two types of reflection can be identified. These are called diffuse and specular reflection and are illustrated in Figure 1.6. Specular reflection occurs when the irregularities of the reflecting surface are small compared to the wavelength of light. The most common example of specular reflection is the image you see in a mirror. The incident rays and the reflected rays are parallel and in the same plane as the normal. Diffuse reflection occurs when the wavelength of the incident light is smaller than the size of the surface irregularities. In this case the incident rays are parallel but because of the roughness of the surface the angle of incidence will vary with each ray. The reflected ray will therefore be reflected at the local angle of incident causing light to be scattered in all directions but still in the plane of incidence. If the surface is rough, then some of the reflected energy may penetrate the surface and, providing the material has an absorption band, the reflected energy will be depleted of that band. Specular and diffuse reflection cannot easily be separated and often reflected light will contain both specular and diffuse components (Schott and Henderson-Sellers, 1984).

A useful measure is the **albedo** or reflection coefficient of a surface. The albedo of a surface is the percentage of **insolation** (solar radiation) incident upon it which is reflected back into space. This definition of albedo implies that it is the percentage of reflection that occurs over all wavelengths of insolation. Sometimes the term albedo is applied to a specific wavelength range

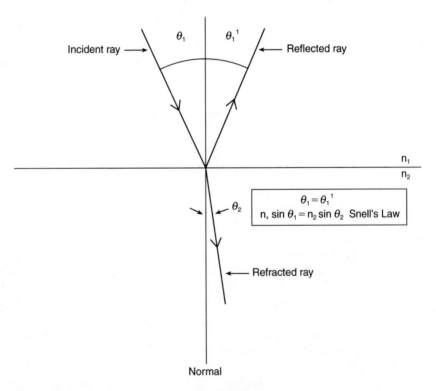

Figure 1.5 Reflection and refraction at a plane surface

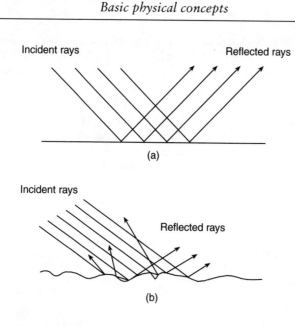

Figure 1.6 (a) Specular-reflection as occurs at a smooth surface (b) Diffuse reflection as occurs at a rough surface

but this is usually made clear from the text. A white body is a perfect reflector and would have an albedo of one. Ice and snow have some of the highest albedos of natural surfaces but will vary greatly depending on the age of the snow, water under the ice and dirt on the surface. Most reflection of insolation from the atmosphere is due to clouds. Different cloud types have different albedos. The planetary albedo refers to the albedo of the Earth and its atmosphere combined.

Scattering

The impression is often gained that **scattering** is little more than a type of reflection. A complete description of scattering, however, is much more than reflection and therefore is beyond the scope of this book. Bohren (1987) says that reflection and refraction are simple forms of scattering and that the former two processes get treated separately in textbooks merely as an historical artefact, because they were described first. With scattering there is an interaction between the radiation and the medium, though absorption need not take place. Due to the three-dimensional nature of electromagnetic waves, this results in light being scattered in all directions. The forward beam of light therefore becomes depleted and of a lower intensity. The remaining light is distributed at other angles. Scattering also results in the polarization of the light. This means that the oscillation of the light wave occurs in one direction only rather than numerous directions. The type of scattering that occurs depends on the wavelength of the incident radiation and also the incident wavelength compared to the size of the scattering particles. Other factors such as the chemical composition of the scatterer also contribute.

When the particles have radii smaller than the wavelength of the incident radiation, Rayleigh scattering occurs. The scattering process is identical in both forward and backward directions and is at a maximum in these directions. The amount of scattering is at a minimum at $90°$ to the incident beam (see Fig. 1.7). In this case scattering varies inversely as the fourth power of the wavelength of the incident radiation. This means that the shorter the wavelength, the greater the scattering. Thus blue light is scattered more than red light. Dry gas molecules are the main Rayleigh scattering agents in the atmosphere and it is scattering by these particles that accounts for the blue colour of the Earth's sky (Bigg, 1997).

Mie scattering takes place as the size of particle approaches the wavelength of the incident radiation In this case more and more light is scattered forward (see Fig. 1.7). It is difficult to define mathematically and so simplifications have to be made by considering atmospheric particles to be spheres or ellipsoids. Water vapour and particles of dust are the main Mie scattering agents in the atmosphere. Individual water molecules are no larger than dry air gases but they tend to coagulate together, creating much larger particles leading to Mie scattering. As particles get even larger, scattering gradually becomes independent of wavelength and direction. This is called Tyndall or non-selective isotropic scattering (*see* Fig. 1.7). Water droplets which are large (5–100 μm) scatter all wavelengths of visible light (0.4–0.7 μm) equally and thus clouds and fog often appear white. Clouds are not always white though and sometimes appear grey or black. This is because of the other process going on, namely absorption.

Absorption

Absorption is where a fraction of the energy passing through a volume element of a substance is absorbed by the substance. The energy is then re-emitted sometimes at longer wavelengths (Schott and Henderson-Sellers, 1984). Absorption is referred to as selective, meaning it only absorbs certain wavelengths. The reason for this lies in how energy is absorbed. So far electromagnetic energy has been discussed in terms of waves, however, electromagnetic energy also exhibits some characteristics of particles. In the particle model, electromagnetic radiation is considered as photons, with discrete packets of energy called quanta. This quantum energy is inversely proportional to wavelength. In atoms the electrons that orbit the atomic nucleus can only do so in pre-defined orbits. The amount of energy required for an electron to change orbit is also a discrete packet of energy i.e. it is quantized. Molecules are formed when atoms bind together with chemical

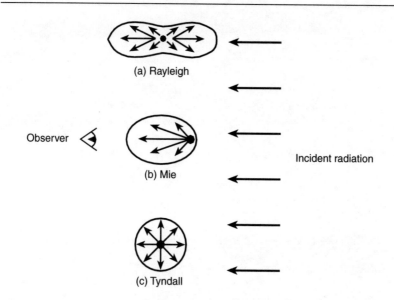

Figure 1.7 (a) Rayleigh scattering (b) Mie scattering (c) Tyndall scattering
Source: After Schott and Henderson-Sellers (1984)

bonds. The structure of the water and carbon dioxide molecule are shown in Figure 1.8. Within a molecule motions can occur, and as movement requires energy, these motions are associated with a change in energy level of the molecule. As nuclei are far larger than electrons, the movement of the nuclei and electrons and the energy transfers associated with those movements can be considered separately.

Molecules can move in four ways: electronic, rotational, vibrational and translational. Electronic motion is concerned with the movement of electrons and is essentially the transfer of electrons between different orbits. Rotational, vibrational and translational movements are all to do with motions of the nuclei. In electronic, rotational and vibrational movements of molecules the amount of energy required is quantized. Therefore, if radiation incident upon a molecule can provide the correct amount of energy, that is if it is of the right wavelength, then this energy will be absorbed by the molecule which will raise it to an excited state. When it falls back to its unexcited state it releases energy again - emission. This is shown in Figure 1.9. There is no need for the molecule to emit the same amount of energy it has absorbed all in one go. It could cascade down various energy levels. You could think of this in terms of throwing a ball up some steps. To reach a particular step you have to impart the right amount of energy to the ball. The higher the step the more energetically you need to throw the ball. This is equivalent to absorption. Emission is equivalent to the ball returning when it may bounce straight back or bounce down several steps before reaching you.

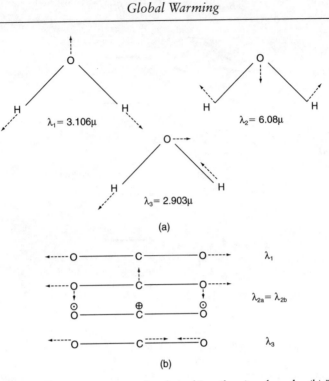

Figure 1.8 (a) The water vapour molecule and its vibrational modes (b) The carbon dioxide molecule and its vibrational modes
Source: Elachi (1987)

Each atom or molecule which makes up a substance will have its own unique energy levels and so will absorb only certain wavelengths. This is why absorption is said to be spectrally selective depending on the absorber. If you look at the spectrum of light that has passed through a substance you can see dark bands or lines where the substance has absorbed those particular wavelengths. These are called absorption lines and they can be used to identify if a substance is present. For example by looking at the absorption lines in the light coming from stars astronomers can tell their chemical make-up. The absorption lines usually appear as narrow bands because environmental factors, such as temperature and pressure, lead to a broadening of the lines.

We can see through various objects such as glass because they do not absorb any radiation at visible wavelengths. However, if we were to look with eyes that could only see in the infra-red then glass would appear opaque as it absorbs wavelengths in the thermal region of the spectrum. In the atmosphere a particular gas molecule can only absorb certain wavelengths of radiation. To summarize in another way, the energy levels of gaseous molecules are well defined and discrete. The related interaction occurs at very specific wavelengths leading to spectral lines. As well as being wavelength dependent the absorption also depends on the amount of the absorber. More absorption will occur with larger amounts of the absorber.

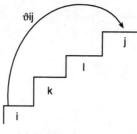

(a) Absorption

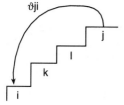
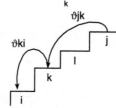
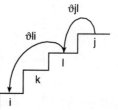

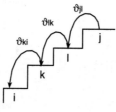

(b) Emission

Figure 1.9 (a) Absorption due to incident radiation of frequency υij (b) Emission which can occur directly υji or by cascading down intermediate levels

Figure 1.10 shows the main absorbing constituents of the atmosphere across the electromagnetic spectrum. Molecular absorbers are active in the far and near infra-red whereas the atomic gases, such as atomic oxygen (O) tend to absorb at shorter wavelengths. The main atmospheric absorbers are oxygen and ozone, carbon dioxide and water vapour. Most important are the 0.01 to 0.3 μm absorption bands of ozone and the 15 μm absorption band of carbon dioxide. The absorption characteristics of the Earth's atmosphere (see Fig. 1.10) show that the atmosphere is nearly transparent to visible wavelengths (0.3 to 0.7 μm) and then there is a series of well-defined absorption bands alternating with transparent regions called windows. The atmosphere is particularly opaque to longwave radiation between 4 to 8 μm. The large transparency in the region 8 to 14 μm is often referred to as the atmospheric window.

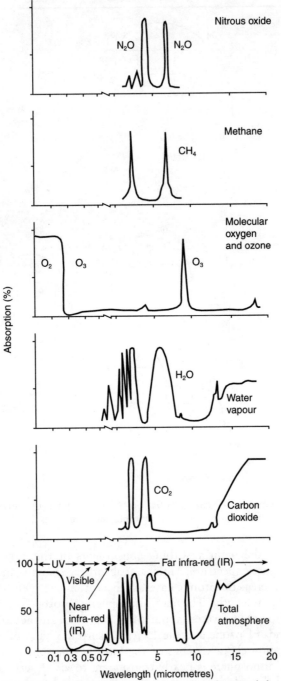

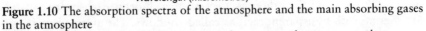

Figure 1.10 The absorption spectra of the atmosphere and the main absorbing gases in the atmosphere
Source: Ahrens (1994). From *Meteorology today: an introduction to weather, climate and the environment,* 5th edition, by C.D. Ahrens © 1994. Reprinted with permission of the Global Rights Group, a division of Thompson Learning.

The Earth's radiation and energy budget

The Earth–atmosphere system derives its energy in the form of radiation from the Sun. A change in the output of the Sun either in intensity or spectral characteristics would affect the energy available to the Earth–atmosphere system and would affect the climate. The total amount of radiation per unit area incident on a surface perpendicular to the Earth at the top of the atmosphere when the Earth is at a mean distance from the Sun is called the solar constant (So). Its present-day value is approximately 1368 Wm^{-2}. Chapter 4 will show that this is anything but constant but for now it will be regarded as such. The amount of energy received at the top of the atmosphere will also vary with the Sun–Earth distance, the altitude of the Sun and the daylength.

First consider radiation at the top of the atmosphere. The Sun provides the main input of energy into the Earth–atmosphere system, and this is balanced on average by emission in the infra-red. Any difference between these two is the net radiation.

net radiation = incoming radiation − outgoing radiation
$$Net = Q - F \tag{1.5}$$

The distribution of incoming energy is not the same as the outgoing energy either in space or time. The oceans and atmosphere store the solar energy and redistribute it. Over a year, however, the global average radiation is such that the incoming solar radiation and the outgoing terrestrial radiation balances, with the net radiation at zero. That is, it will reach an equilibrium so the Earth neither warms up nor cools down. At the top of the atmosphere, the incoming radiation averaged over the whole globe is

$$\text{incident radiation} = So\pi r^2 / 4\pi r^2 = So/4 \tag{1.6}$$

A proportion of the incident radiation is reflected away by clouds and other highly reflecting features such as snow and ice. The portion of incoming radiation which will be absorbed and therefore play a part in the climate system is given by

$$Q = So(1-a)/4 \tag{1.7}$$

where a is the planetary effective albedo of the Earth, approximately 0.3.

When considering the outgoing terrestrial radiation the modified Stefan-Boltzmann law needs to be used as the Earth is not a perfect black body. The emissivity of the Earth is usually given as 0.95. It is now possible to go back to equation (1.5) and, with the net radiation set to zero, write it as

$$Q = F$$
$$So(1-a)/4 = \varepsilon\sigma T^4 \tag{1.8}$$

This equation can be rearranged:

$$T^4 = So(1-a)/4\varepsilon\sigma \tag{1.9}$$

The equation can now be solved for temperature. The temperature of the Earth, as given by equation (1.9) is 255 K which is $-16°C$. This is the planetary effective temperature of the Earth. The temperature as seen by a satellite. However, this is obviously not the same as the surface temperature which is a balmy 288 K. Why is there a temperature difference of some 33 K between the top and bottom of the atmosphere? The answer to this lies in the composition of the atmosphere and the process of absorption.

Figure 1.11 is a representation of the Earth–atmosphere radiation and energy budget. In this representation the incoming solar radiation at the top of the atmosphere is said to be 100 units. Of those 100 units some 30 units are reflected back into space. Most of this, around 26 units, is due to atmospheric reflection and scattering predominantly by clouds. A further 4 units are reflected by the Earth's surface. Around 19 units will be absorbed by the atmosphere, the remaining 51 units are then absorbed by the Earth's surface. At the surface other types of energy transfers take place, namely latent heat and sensible heat transfers. Some 23 units are used to evaporate water and a further 7 units are lost through conduction and convection. This leaves 21 units. At all times the energy must balance for equilibrium to be maintained, so one might expect that as the surface is gaining 21 units it must lose 21 units in infra-red radiation but looking at Figure 1.11 shows that the amount of outgoing radiation units from the surface is 117 units.

The surface temperature is 33 K greater than the planetary effective temperature at the top of the atmosphere. The question is still, what causes this difference? Consider what happens to those 117 units. Of these only 6 units escape directly into space. The remaining 111 units are absorbed by the atmosphere principally by water vapour and other greenhouse gases such as carbon dioxide. The atmosphere will then re-emit this energy and in all directions, so not only will some be lost to space, 64 units, but most, 96 units, will go to warm the Earth's surface. This is the atmospheric greenhouse effect. It is this effect which allows the Earth's surface to radiate 117 units and so achieve a higher surface temperature than the planetary effective temperature. To summarize the radiation budget, on the whole, solar radiation gets through the atmosphere to warm the Earth's surface whereas terrestrial radiation is trapped. The atmospheric constituents absorb the outgoing radiation then reradiate it, some back to the surface. The result is a surface temperature of 288 K not the 255 K viewed from space. In Figure 1.11 it is important to note that everywhere there is an energy balance. At the Earth's surface the 147 units lost from the surface by infra-red radiation and sensible and latent heat transfers are balanced by a 147 unit gain through solar and infra-red radiation. In the atmosphere the 160 unit gain through sensible and latent heat transfers and solar and infra-red radiation is offset by a 160 unit loss by solar and infra-red radiation. At the top of the atmosphere the 70 unit gain from solar radiation is balanced by a 70 unit loss by infra-red radiation.

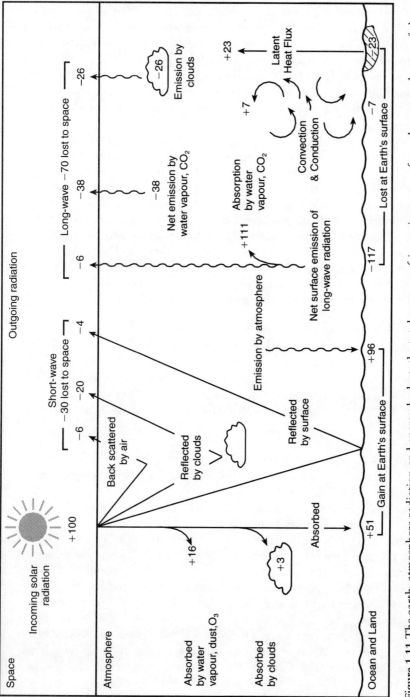

Figure 1.11 The earth–atmosphere radiation and energy budget: the total amount of incoming energy from the sun at the top of the atmosphere is represented as 100 units

Summary

From the concepts introduced in this chapter you should be able to see that if human activities change the composition of the atmosphere then potentially it can affect the amount and spectral composition of incoming solar radiation and similarly affect the outgoing radiation. In perturbing the balance of incoming and outgoing radiation, and the amount of energy available to the Earth–atmosphere system the surface temperature and therefore the climate can be changed.

Figure 1.12 shows the standard diagram of the climate system, its components and processes. The components of the climate system atmosphere, hydrosphere, cryosphere and biosphere all interact with one another by a vast number of different physical and chemical processes. For instance, physical processes such as freezing, winds, chemical processes such as methane release from swamps, and radiative processes such as absorption and scattering. The climate is the net balance between all these components and processes. The next chapter will introduce the various components of the climate system and consider how they interact. It is within these interactions that the complexity of the climate system lies. It is this what makes the question of climate change and global warming such a challenging one.

References

Ahrens, C.D. 1994: *Meteorology today: an introduction to weather, climate and the environment*, fifth edition. Minneapolis/St. Paul: West.

Bigg, G.R. 1997: Light in the atmosphere: part 1 - why the sky is blue. *Weather* 52, 72–7.

Bohren, C.F. 1987: *Clouds in a glass of beer: simple experiments in atmospheric physics*. New York: John Wiley and Sons.

Elachi, C. 1987: *Introduction to the physics and techniques of remote sensing.* New York: John Wiley and Sons.

Houghton, J.T., Jenkins, G.J. and Ephraums, J.J. (eds) 1990: *Climate change: the IPCC scientific assessment.* Cambridge: Cambridge University Press.

Lockwood, J.G. 1979: *Causes of climate.* London: Edward Arnold.

Schott, J.R. and Henderson-Sellers, A. 1984: Radiation, the atmosphere and satellite sensors. In A. Henderson-Sellers (ed.) *Satellite sensing of a cloudy atmosphere: observing the third planet.* London: Taylor and Francis.

Sears, F.W., Zemansky, M.W. and Young, H.D. 1978: *College physics*, fourth edition. Reading, MA: Addison-Wesley.

WMO 1975: *The physical basis of climate and climate modelling.* GARP publication series 16, Report of the international study conference in Stockholm, 29 July–10 August 1974. Geneva: World Meteorological Office.

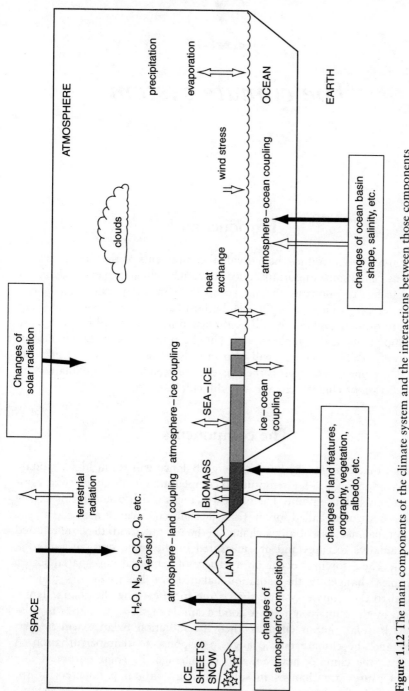

Figure 1.12 The main components of the climate system and the interactions between those components
Source: WMO (1975)

|2|

The climate system

Introduction

This chapter will introduce the individual components of the climate system as well as define some important terms concerning climate change. Chapter 1 listed the five components that make up the climate system; atmosphere, hydrosphere, cryosphere, biosphere and lithosphere. The processes that occur in the atmosphere and ocean will be discussed in some detail because these two components of the climate system have been the focus of climate modelling attention. The chapter will also introduce the concept of feedbacks within the climate system. It is the existence of feedbacks that explains why the prediction of climatic change is so difficult.

The components

Anything that changes the Earth's energy balance will result in the climate system readjusting itself, attempting to compensate for the change and coming to a new balance point. The natural changes which occur in this energy balance are usually called **forcings**. As these forcings cause a change in the radiation balance, essentially a change in the net radiation, they are referred to as **radiative forcings** and are measured in the same units as radiation (Wm^{-2}). These forcings occur on a variety of temporal and spatial scales. Sometimes changes in the radiation balance are due to factors that are external to the climate system; that is the force causing the change is not affected by the climate system. A good example of this is a change in solar output. If solar output changes, then the radiation balance will be perturbed and the climate may change in response to that perturbation. A change in the climate, however, will not change the solar output. Other external causes are changes in solar intensity, orbital parameters of the Earth, and the Earth's rotation rate. External terrestrial forcings are changes in the atmospheric composition due to geophysical processes (Peixoto and Oort, 1992). These include weathering and volcanic activity, variations in land surface, long-term tectonic activity such as mountain building, continental drift and polar wandering. Other forcings are internal to the climate system and are affected by the climate. For example, increased snow cover can change the climate by reflecting away more solar energy. However, snow cover is itself determined by the climate. Internal causes are associated with many **feedbacks** and interactions between the components of the climate system. Examples are changes in the size of the ice sheet, and changes in sea surface temperatures, anomalies of which may feed back into the system and cause more change. The distinction between external and internal forcings is not always obvious. The mechanisms responsible for changes in the radiative forcings are considered in more detail when the various climate changes are discussed. In both the examples of external and internal forcings it is assumed that a radiative forcing directly relates to a change in climate. In reality, the response of the climate to a forcing is far more complex than this.

When studying the climate system and climate change both spatial and temporal scales are very important. Exchanges of mass, momentum and energy take place at many different spatial scales; sometimes it is possible to ignore certain spatial scales and at other times not. Time is very important because changes in the energy balance and the consequent response of the climate system are time dependent. Each component of the climate system reacts to such changes on a different time-scale. The **response** or **relaxation time** is the time taken for a component of the climate system to return to equilibrium after a change in the forcing conditions.

The atmosphere responds very quickly to changes in forcing, such as an increase in the amount of solar radiation. It warms up and quickly reaches a stable temperature. In weather forecasting meteorologists are typically only interested in 10 days ahead. The only component of the climate system to have a response time that short is the atmosphere. So meteorologists can afford to restrict their studies to the atmosphere and hence the definition of weather given in Chapter 1. Looking beyond short-range and medium-range weather forecasting requires longer time-scales and some aspects of the upper ocean can respond and these have to be included. The longer the time-scale of the study, the greater the number of components of the climate system that have to be considered. Figure 2.1 outlines the time-scales and events of various processes and external forcings. In Chapter 1 when systems were discussed it was pointed out that each component of the climate system was a system within itself. The climate system is thus made up of sub-systems that are continually responding to changes. However, some parts of the system will respond faster, leading others which lag behind (Peixoto and Oort, 1992).

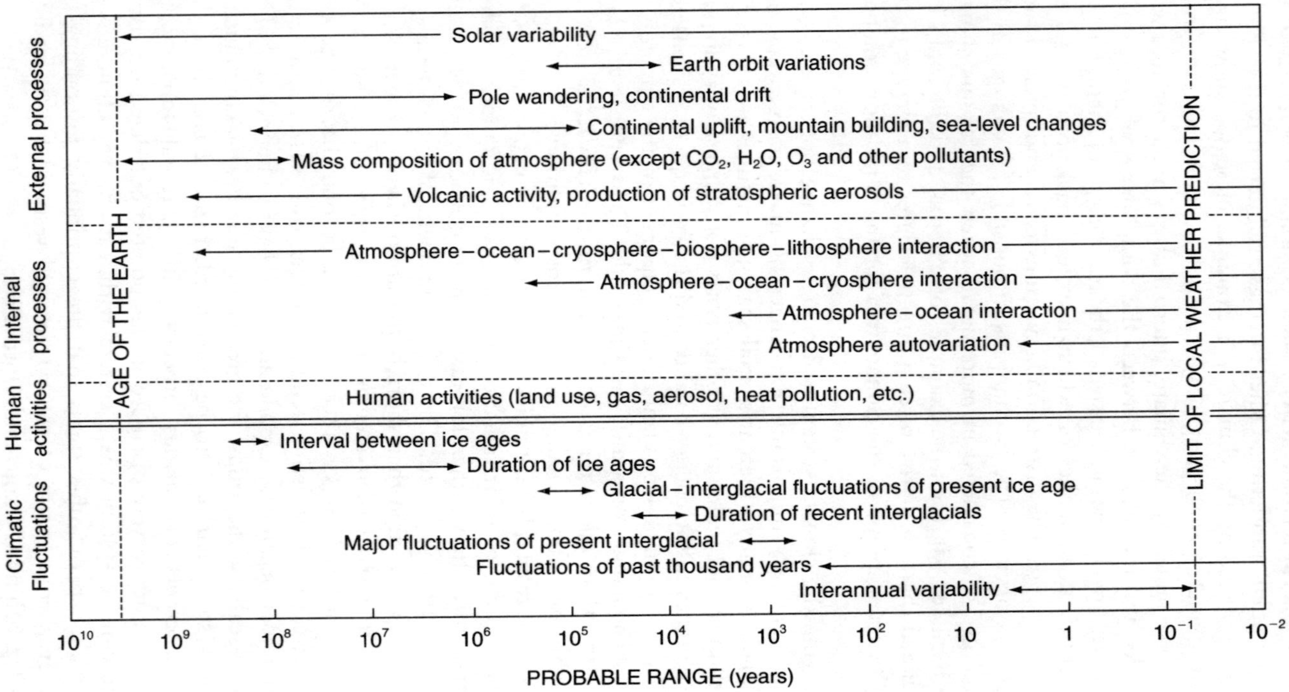

Figure 2.1 Time-scales for individual components of the climate system and some external forcing factors
Source: Peixoto and Oort (1992)

The atmosphere

The atmosphere is the most variable component of the climate system and its composition is given in Table 2.1. Table 2.1 shows that the atmosphere is predominantly nitrogen and oxygen with a few other gases, notably carbon dioxide and water vapour. The amount of water vapour in the atmosphere is highly variable. The vertical temperature profile of the Earth's atmosphere, as shown in Figure 2.2, reveals distinct layers (or spheres) in the atmosphere defined by the temperature structure. Some 80 per cent by mass of the atmosphere is in the troposphere, the lowest layer (10 – 15 km) and it is here that most weather occurs. In the troposphere (lower atmosphere) the temperature decreases with height. This allows convective motion. **Troposphere** literally means turning sphere and refers to this constant overturning of the atmosphere by convection. When you look at a picture of the Earth, the atmosphere's most visible feature is cloud. Water vapour and clouds are confined mostly to the troposphere as the tropopause does not allow water vapour to penetrate into the next layer, the **stratosphere**. This is because the tropopause is a region of almost no temperature change with height that effectively stops convective activity and acts as a lid. The altitude of the tropopause varies with season and latitude. It occurs at around 10 km altitude at the poles, increasing to 15 km at the equator. The height difference from equator to pole is due to the greater amount of solar heating received by the equatorial regions.

Table 2.1 Atmospheric composition: major constituents

Constituent gas	Chemical formula	Proportion (%)
Nitrogen	N_2	78.1
Oxygen	O_2	20.9
Argon	Ar	0.93
Water Vapour	H_2O	0.1-1
Carbon Dioxide	CO_2	0.0355
Methane	CH_4	0.000172
Nitric Oxide	N_2O	0.000031
Ozone	O_3	Variable around 0.000005

Source: Bigg (1996)

The next two layers, the stratosphere and **mesosphere**, form the middle atmosphere. The stratosphere occurs above the tropopause and is a layer where the temperature increases with height. The temperature increases with height because the stratosphere contains the **ozone layer**. The previous chapter showed that ozone had strong absorption bands at ultra-violet wavelengths. Therefore ozone acts as an effective absorber for these wavelengths which is also the most carcinogenic part of the solar spectrum (Solomon, 1988). The ozone layer thus absorbs energy warming the stratosphere. It is not exactly a layer but is a region, the exact location of which varies from season to season and with latitude. The region occurs approximately at an alti-

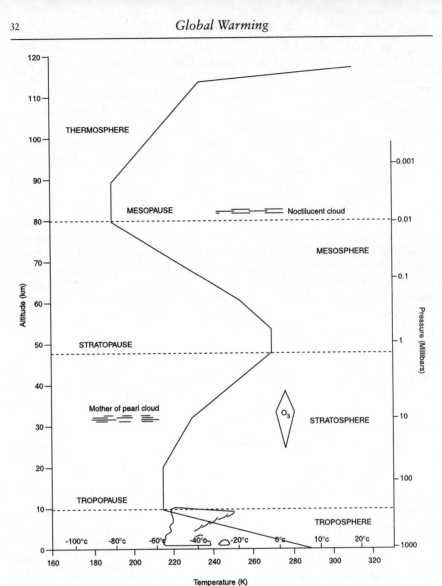

Figure 2.2 Vertical temperature structure of the Earth's atmosphere
Source: After McIlveen (1992)

tude of 15 to 40 km with a maximum concentration occurring at 25 km. The ozone layer therefore screens the Earth's surface from these harmful rays. Since the 1970s there has been growing concern that synthetic compounds (**chlorofluorocarbons**) and other compounds that are being increased by human activity may be depleting the ozone layer. This may result in increased levels of ultra-violet radiation at the Earth's surface and consequently increased health risks. These concerns over the ozone layer placed it at the centre of a major environmental debate and international legislation (Drake, 1995). The stratosphere is topped off with the stratopause, which like the tropopause, is an area of almost constant temperature with height. Above the stratopause lies the mesosphere. This is a layer of decreasing temperature with height. This stretches to 80 km and this is where the mesopause occurs. Above the mesopause lies the upper atmosphere. The thermosphere is part of this and is a region of increasing temperature with height. Here the atmosphere becomes increasingly rarefied. It is changes in the troposphere, the lower atmosphere, that mostly affect weather and climate. However, some changes taking place in the middle atmosphere, particularly the stratosphere, may have the potential to alter the lower atmosphere. This is because weather and climate are determined by the general circulation of the atmosphere which in turn is determined by temperature variations in the atmosphere. For example, decreases in the ozone layer will result in less ultra-violet radiation being trapped in the stratosphere and the stratosphere will cool. Such a change in the temperature structure of the stratosphere may then feed into the troposphere altering the general circulation. Whether changes in the upper atmosphere affect the climate is still not clear (NRC, 1994).

Although it may seem to be vast, the atmosphere is a very thin film. If compressed to surface temperature and pressure conditions it would be just 8 km thick. That is about 0.1 per cent of the Earth's surface (Harries, 1990). The atmosphere is a fluid and has fluid properties just like water. This has important implications for its motion across the Earth. The Earth may appear stationary but it is revolving on its axis at a speed of 465 ms^{-1} at the equator, as well as revolving around the Sun at 29.8 kms^{-1}. The result of this circular motion is that the atmosphere is subject to forces that affect its movement. One of the most important of these is the **coriolis force**. The coriolis force is best illustrated by an example in just two dimensions. Imagine that you are icing a cake that is on a revolving turntable. You decide to ice in a straight line from the centre of the cake to the outside. However, once you stop the turntable you will find that you have made a curved line. This is because as you moved from the centre the cake was moving underneath. A similar thing happens in the atmosphere, as a parcel of air moves over the rotating Earth it will appear to take a curved path rather than a straight line. The coriolis force has acted on it deflecting the parcel from moving in a straight line. The coriolis force is dependent on latitude and is zero at the equator and greatest at the poles.

Chapter 1 discussed how there is an energy imbalance between the equator and the poles. The energy and consequent temperature imbalance between the equator and poles provide the energy to drive the atmospheric circulation. Most of the motion is in the horizontal plane across the surface of the Earth although obviously there is some vertical motion. The oceans also transfer some heat, perhaps as much as one third. The remaining energy is transferred by the atmosphere. The general circulation of the atmosphere (and oceans) does not just provide an energy transfer; it must also maintain the global water balance, the atmospheric mass and the angular momentum of the atmosphere. The control that the general circulation of the atmosphere has over the global precipitation regime is great and, as the next section shows, the two are closely linked.

THE GENERAL CIRCULATION

The **general circulation** of the atmosphere is a term used to describe the average global wind patterns, both at the Earth's surface and in the upper atmosphere. Wind is usually defined from the direction it has come from, so a westerly wind is one which flows from west to east. **Zonal** components refer to those horizontal winds travelling east or west, parallel to lines of latitude. **Meridional** components are those flowing north and south, parallel to lines of longitude. The troposphere is really the main area of interest for this is where the bulk of the atmospheric motions important to weather and climate take place. Some features of the atmospheric general circulation are very long-lived and cover large areas of the globe. Other characteristics such as depressions (**cyclones**) and **anti-cyclones** are more ephemeral. It is through these long- and short-lived wind regimes that the mechanical transfer of heat takes place.

Either side of the equator are two thermally driven convective cells often called **Hadley cells** after the eighteenth-century meteorologist who first described them. The Hadley cell circulation is meridional (north–south) and is the dominant flow pattern for the region (*see* Fig. 2.3), thus allowing a poleward transfer of energy. There is some zonal (east–west) flow from air rising over Southern Asia (also South America and Indonesia) in the western tropical Pacific and descending air in the east. This is known as the Walker Circulation. At the equator the surface is heated by the Sun and provides the energy for the lower atmosphere and warm moist air can rise. The vertical motion creates giant cumulonimbus clouds. These convective clouds provide latent heat to the ascending air which provides further energy for the convective motion. The convective clouds also provide large amounts of precipitation. This region of strong convective motion is frequently referred to as the **Intertropical Convergence Zone** (ITCZ). The **Equatorial Trough** is a broad zonal trough of surface low pressure where the surface winds, called the **trade winds**, converge. On a satellite photograph of the Earth, the ITCZ is marked by a band of thick clouds. It is not a continuous band of cloud, however, there are preferred areas of convergence. One of these occurs over the South Pacific and is called the **South Pacific Convergence Zone** (SPCZ). This area of convergence extends from the Indonesian region south-eastwards into central South Pacific. While the ITCZ and Equatorial Trough may often co-exist in the same place they are not one in the same thing. The Equatorial Trough is more strongly connected to the thermal lows caused by solar heating. The area of maximum cloudiness and the ITCZ are sometimes displaced away from the trough (Barry and Chorley, 1992).

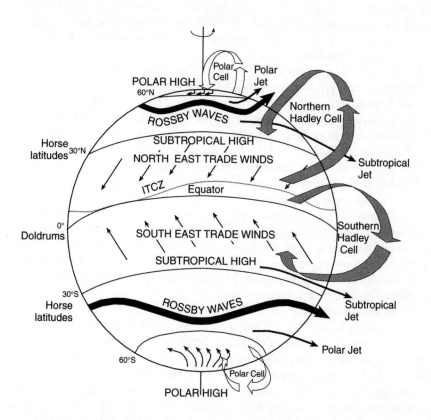

Figure 2.3 The present-day representation of the general circulation of the Earth's atmosphere

The upward convective motion is stopped by the tropopause and the air is forced to move polewards. Thus at the equator the convergence at the surface is balanced by an outflow of air, divergence, at the tropopause. The coriolis force ensures that the north – south flow is deflected creating a westerly flow which reaches a peak at 30° latitude and results in a **jet stream**. The air high up in the troposphere cools by radiation. At about 30° north and south convergence aloft occurs and a surface high pressure zone is created, known as the sub-tropical highs (or anti-cyclones). The sub-tropical highs are associated with the world's desert regions. Maps of precipitation show that areas of low precipitation also occur over the oceans at the sub-tropical latitudes. The lack of precipitation is due to warm dry air subsiding towards the surface making it almost impossible for clouds to form and hence precipitation to occur (Lockwood, 1979). At the surface some of the air flows back to the equator as the trade winds. These winds blow equatorward from the east, roughly north–east in the northern hemisphere and south–east in the southern hemisphere. The trade winds do not blow directly north–south due to the coriolis force. In the days when world trade relied on sailing ships these winds were very important as they provided ships with a remarkably constant wind speed and direction.

The two Hadley Cells either side of the equator, with the trade winds meeting at the equator is only found in spring and autumn. Usually the ITCZ is displaced into the summer hemisphere. The movement of the ITCZ is more pronounced over the continents and it provides the wet and dry seasons of the equatorial regions. Many people rely on this annual movement for their water and their crops. If the ITCZ fails to move sufficiently it can bring severe drought and hardship to these regions (Kemp, 1990). In the summer hemisphere there is often a band of light variable winds separating the two trade systems particularly over continental areas. These are known as the Doldrums and were areas where sail-powered trading ships could be stranded. At the poleward edge of the Trades over the oceans there is another area of light variable winds, which alter considerably with season, associated with the sub-tropical high pressure systems. These are known as the Horse Latitudes as trading ships stuck here were often reduced to either eating their horses or throwing them overboard.

Monsoon regimes, which are found at low latitudes, also have a wet and dry season. Failure of the wet season can lead to drought, while excessive rains can lead to flooding (O'Hare, 1997a and b). Although there are links, these seasons are not due to the movement of the ITCZ. Monsoon is a term first used by Arabic writers in the middle of the tenth century. It describes a seasonal change in the surface wind pattern over southern Asia. The region has twice yearly changes, with the wind reversing direction in winter and summer. The term monsoon is now applied to other areas of the globe, such as in Africa, which also exhibits a similar seasonal wind cycle (O'Hare, 1997a). The climate of the Indian sub-continent remains as the classic example of a monsoon regime. The wet season is hot with humid winds. The dry season is cooler with winds of lower humidity. The classical theory of the monsoon circulation describes it as a large-scale onshore breeze (O'Hare, 1997a). In spring the continental interior warms up creating a thermal low. Warm moist air is then drawn in from off the oceans into the continental interior. Here the air rises releasing precipitation and latent heat which helps to maintain system. Over the oceans there is cool descending air. While this classical theory explains many of the observed features, it cannot explain them all. In fact, the monsoon is as much governed by the movement of the sub-tropical jet stream as by the creation of thermal lows. It is the transition of the sub-tropical jet stream from the south of the Himalayan plateau to the north in late May which allows the monsoon rains to burst into the interior. The Asiatic monsoon is a result of complex interactions between tropical and mid-latitude upper circulations and the distribution of land, water and topography (O'Hare, 1997a). The strength of the monsoon is also strongly linked to the El Niño Southern Oscillation (ENSO).

Some of the air from the descending Hadley cell moves polewards, rather than equatorwards. As it moves polewards it is deflected by the coriolis force. At these mid-latitudes the coriolis force is larger than in the subtropics and so provides a westerly wind. Unlike the trade winds, however, the flow is far from reliable. The winds are constantly being broken up by travelling low and high pressure systems (cyclones and anti-cyclones). In the northern hemisphere the greater land mass also disturbs the westerly flow. In the southern hemisphere the westerlies are more marked as the oceans do not allow the formation of stationary pressure cells. The warm air travelling poleward from the equator eventually encounters cold air coming down from the poles, at around 60° latitude. The two air masses do not interact easily and form the polar front, which nearly encircles the hemisphere. It marks a temperature discontinuity and continues upwards into the troposphere creating a frontal surface. There is an increase in wind speed with height and a jet stream is formed at or just below the tropopause.

The mid-latitudes are an area of particular complexity. Unlike the equator where the Hadley cell ensures a strong meridional flow, in the mid-latitudes this is very weak. Most of the flow is zonal but energy is still transported from lower to higher latitudes. This is achieved not by a convective cell but by waves which propagate around the globe. These are called **Rossby waves**, after the meteorologist who determined them mathematically. They affect the whole depth of the troposphere over most of the mid-latitudes and are also called planetary waves. Rossby waves occur because of the spinning of the Earth and the atmosphere, called the vorticity, and the coriolis force. The Rossby or planetary waves have a pseudo-cyclic change in amplitude. As the amplitude of the wave changes there is an index cycle which is shown in Figure 2.4. This index is given by the zonally averaged pressure difference between two latitude circles. In the northern hemisphere these are 35° and 55° latitude which usually define the main Rossby flow. When the pressure difference between these two latitude circles is large then there is a high index. This results in a strong zonal flow and smooth Rossby waves which are often north of their mean position. As the pressure difference decreases so the index decreases but the amplitude of the waves and their velocity increase. At this time the zonal westerlies are frequently south of their mean position and may break-up forming distinct high and low pressure centres. Next the amplitude may decrease and a return to a high index occur. Alternatively, cut-off can occur. This results in relatively stationary atmospheric features: a blocking anticyclone at high latitudes, and a non-frontal depression at low latitudes. In the northern hemisphere the cycle is particularly pronounced during February and March. The time it takes to complete a cycle is very variable ranging from 3 to 8 weeks and each stage of the cycle also varies in length. All these features can be observed in the upper troposphere where a strong westerly flow dominates (Barry and Chorley, 1992; Henderson-Sellers and Robinson, 1986).

Although some of the features of the index cycle, such as a blocking anticyclone, may appear on surface charts, this is not usually the case. Surface

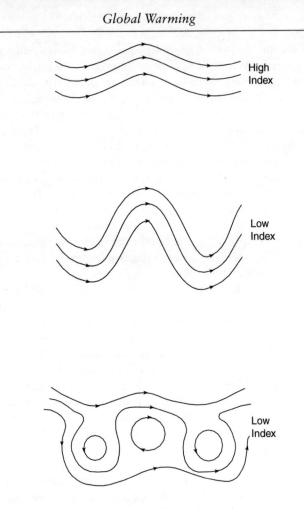

Figure 2.4 Rossby wave index cycle

depressions, low pressure areas, occur at the mid-latitudes because there is a distinct poleward temperature gradient. By considering this temperature gradient in conjunction with the vorticity and coriolis force it can be shown mathematically that the mid-latitudes are areas of great instability. The atmospheric conditions in the mid-latitudes are such that cyclogensis, the formation of depression systems, is bound to occur. Depression formation is associated with a strong zonal flow in the upper westerlies. A time of high index flow is, in general, a time of deep depressions. It is upper air divergence caused by the poleward moving portion of the Rossby wave that leads to surface convergence and the formation of depression systems. The complete picture is far more complex than this with the Rossby waves, polar front and jet stream all interacting with the surface features; each affecting the other. The surface low pressure systems provide the mid-latitudes with its rapidly changing weather. They also transport energy poleward, supplementing the Rossby wave distribution of energy (Henderson-Sellers and Robinson, 1986).

At the poles there is a theoretical polar cell which is similar to the Hadley cell. The coriolis force, however, is large so the surface flow is deflected into an almost zonal easterly flow, though these are weak over the Arctic. Above the surface westerlies dominate. Consequently there is little meridional flow and the polar cell is not well developed. Instead the motion and energy transfer might be more akin to those in the mid-latitudes which are essentially zonal in nature (Kemp, 1990). The poles are small regions and are the areas to which energy is being transported, so they do not play a great role in energy transferral.

The hydrosphere

The hydrosphere includes all water in the liquid phase. It consists primarily of the oceans but it includes lakes and rivers. Clouds, however, are usually considered as part of the atmosphere. The oceanic general circulation, just like that of the atmosphere, redistributes energy from the equator to the poles. Although there are many similarities between the oceanic and the atmospheric circulation there are also many differences. The ocean is constrained by the oceanic basins and so it is not as free to move as the atmosphere. The main oceanic basins are the Pacific, Atlantic and Indian Oceans. The oceans do not respond as quickly to changes in radiative forcing as the atmosphere, and have a response time in the upper circulation of weeks to months and in the deep circulation of centuries to millennia. Of great importance to the climate system are the interactions between the atmosphere and oceans. They are said to be strongly coupled, which means that changes in one will result in changes in the other. An example of this is the upper oceanic circulation which is primarily driven by the surface winds. The oceans also exchange carbon dioxide and other important gases and **aerosols**, with the atmosphere. The horizontal and vertical movements of the atmosphere and oceans control the energy budgets of all locations.

OCEANIC COMPOSITION AND STRUCTURE

Table 2.2 shows the composition of standard ocean water. Density is the mass per unit volume of a substance and pure water has a density of 1.00 g/cm^3 at 4°C. This is the temperature at which water is most dense. This is an important point because as the temperature of fresh water decreases and drops below 4°C , it is lighter, less dense, and so floats. By adding salts to water the properties of the water are changed. Sea water is more dense than fresh water, typically it has a density of 1.028 g/cm^3. If sea water contains less than 24.7 g of salt per kilogramme of water it acts like freshwater, so it reaches its maximum density before it freezes. At 24.7 g/kg the freezing point and maximum density both occur at −1.332°C. At greater than 24.7 g/kg as the temperature of the salt water decreases, its density also increases until freezing occurs. As open ocean water has a salt content of 35 g/kg, the density of the sea water will increase as cooling occurs. This then allows the cooled surface water to sink (Duxbury and Duxbury, 1989). This behaviour is very important for the oceanic circulation and has important implications for the climate system which will be described in later chapters.

Table 2.2 Standard ocean composition

Constituent ion	Chemical symbol	Mass of ion in grams per 1kg of sea water
Chloride	Cl^-	19.350
Sodium	Na^+	10.760
Sulphate	SO_4^{2-}	2.712
Magnesium	Mg^{2+}	1.294
Calcium	Ca^{2+}	0.412
Potassium	K^+	0.399
Bicarbonate/Carbonate	HCO_3^-	0.145

Source: Bigg (1996)

Thus about 3.5 per cent by volume of sea water is dissolved salts. The salts listed in Table 2.2 comprise 99.9 per cent of all dissolved salts in the ocean. The remaining 0.1 per cent are various stable elements. The salts are present because of the ease with which ionic compounds (those which have a positive or negative charge) are dissolved in water. The salts are brought to the sea by rivers or come from land weathering, volcanic and hot spring discharges. This input of salts is balanced by a series of sinks in the form of sedimentation on the ocean floor and continental shelves and transfers to the atmosphere and biosphere. The total salt content of the oceans has been in equilibrium for perhaps 100×10^6 years. This time period is thought to be the time necessary for life to evolve (Wells, 1986).

In 1872 the *Challenger* expedition was mounted under the auspices of the British Royal Society. HMS *Challenger* collected a wide variety of oceanographic data in its 5-year mission. The data took 20 years to be analysed and the voyage is usually credited as being the start of modern oceanography (Duxbury and Duxbury, 1989). William Dittmar studied the oceanic constituents from samples collected by HMS *Challenger*. He was able to conclude that the major constituents of sea water were present in constant proportions to one another regardless of salinity. For example there will always be twice as much sulphate as magnesium (see Table 2.2). Therefore in order to measure the salinity of sea water oceanographers need only measure one of the constituents. Chlorine, being one of the easiest constituents to measure chemically, is frequently used to measure **salinity**. Chlorinity (ClO/OO) is the amount of chlorine in grams per 1 kg of sea water, parts per thousand (0/00), where both the bromine and iodine have been replaced by chlorine.

Thus:

$$S(0/00) = 1.80655 \, Cl(0/00) \tag{2.1}$$

Measurements of salinity are now made by assessing the electrical conductivity of the sea water and a scale based on this has been devised. The average salinity of the oceans is approximately 35 practical salinity units (psu). These dimensionless units are virtually identical to the measure of parts per thousand (Bigg, 1996). In coastal areas of high precipitation and river inflow the salinity is below average. In regions of high evaporation and low precipitation high salinities occur such as in the subtropics and landlocked oceans such as the Red Sea and Mediterranean.

THE GASES

An important form of air–sea interaction is the movement of gases between the sea and the atmosphere. Mixing processes and currents distribute the dissolved gases throughout the depths of the ocean. In both the atmosphere and ocean the dominant gases are nitrogen, oxygen and carbon dioxide. For most gases the ocean only contains a small percentage of that contained in the atmosphere, the major exception being carbon dioxide. A large amount of carbon dioxide is contained in the ocean and ultimately it is the ocean that determines the equilibrium level of carbon dioxide in the atmosphere. The ocean acts as a store for carbon through the carbon dioxide cycle (Stanton, 1991). The dissolved carbon dioxide also acts as a buffer to prevent sudden changes in the acidity or alkalinity of the sea. Maintaining a constant pH value is important for many sea-dwelling organisms (Duxbury and Duxbury, 1989). The rate of uptake of carbon dioxide by the oceans is controlled by the water temperature, the acidity, the salinity, the chemistry of the ions, biological processes as well as mixing and circulation patterns. The chemical movement from the atmosphere and to the oceans is given by

$$CO_2 + H_2O \Leftrightarrow H_2CO_3 \Leftrightarrow HCO_3^- + H^+ \Leftrightarrow CO_3^{2-} + 2H^+$$

Key:
H_2CO_3 carbonic acid
HCO_3^- bicarbonate
H^+ hydrogen ion
CO_3^{2-} carbonate

This carbon dioxide cycle forms part of the larger carbon cycle as shown in Figure 2.5. The carbon cycle explains how carbon in various forms is transferred from one part of the climate system to another. Typically gases are produced by a source then they are sequestered (taken up) by a sink which effectively removes them from playing a further part in the climate system. In contrast, carbon dioxide, because it forms part of the carbon cycle, has no real sinks but the various components of the climate system act as reservoirs or temporary stores of carbon. **Photosynthesis** is the process by which plants

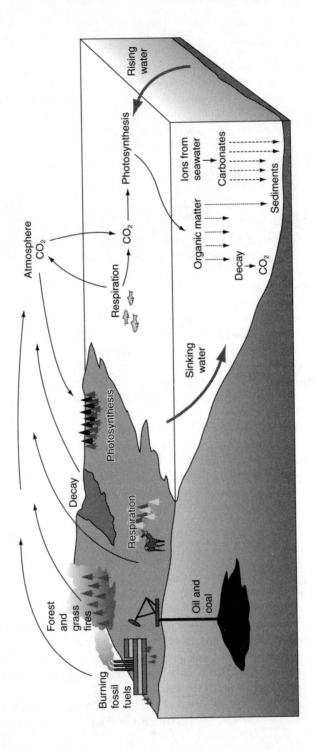

Figure 2.5 The global carbon cycle. Reproduced with permission from the McGraw-Hill Companies: Duxbury, A.C. and A.B. Duxbury 1989: *An introduction to the world's oceans.* Dubuque: Wm. C. Crown.

and algae take up carbon dioxide and water and convert these into carbohydrates and oxygen using the energy from sunlight. Most of the oxygen is then released into the atmosphere. The carbon dioxide taken up by marine organisms is transferred into the deep ocean when the organic matter decays and sinks to the ocean floor. This is often in the form of calcium carbonate ($CaCO_3$) sediments, effectively the skeletal remains of the marine organism. Vast amounts of carbon dioxide are trapped in marine sediments in this way. This movement of carbon down into the deep ocean is often referred to as the **biological pump** because biological processes control the transfer. Oxygen is taken up by organisms during respiration and carbon dioxide and water vapour are released back into the atmosphere. The carbohydrate is also converted back to carbon dioxide and water vapour during respiration. A certain amount of atmospheric oxygen is also taken up during the decay of organic matter and again results in the release of carbon dioxide and water. Oceanic circulation patterns also affect carbon dioxide in the atmosphere. Cold, carbon dioxide enriched water sinks in the polar regions locking carbon dioxide away in the deep ocean. Warm upwelling water in the tropics and eastern boundaries of oceans returns some carbon dioxide to the atmosphere. There is also a continuous exchange of carbon dioxide between the atmosphere and the ocean of intermediate depths.

Through the carbon dioxide cycle the oceans also help to regulate the oxygen content of the Earth's atmosphere. The conversion of marine organic matter into sediments means that it does not decay and take up oxygen. Therefore an imbalance results between the production of oxygen through photosynthesis and its removal, resulting in an annual excess of atmospheric oxygen of 300 million tonnes. Oxygen does not build up in the atmosphere because the process of weathering and the oxidation of rock material removes the surplus (Duxbury and Duxbury, 1989).

Within the carbon cycle there are some temporary stores which by human standards occur over a very long time period, such as the sedimentary part of the carbon cycle. On the time-scales that are important to global warming, typically tens to hundreds of years, these cease to have any significance as they take too long to occur. Carbon dioxide exchanges between the atmosphere, the biosphere and the ocean are the most important. The excess carbon dioxide from anthropogenic emissions is initially released into the atmosphere. It then takes about four years for any one molecule of carbon dioxide to be transferred into the ocean or the biosphere. The excess carbon dioxide emissions become part of the carbon cycle. However, just like the climate system the cycle has to adjust to the increased carbon dioxide levels. It takes time to come to equilibrium once a change has occurred. Also, as with the climate system, the time it takes to re-equilibriate depends on the slowest component and in this case it is the exchange of carbon that occurs from the ocean surface to the deep ocean and that can take up to 200 years.

TEMPERATURE AND SALT STRUCTURE OF THE OCEANS

The main variables describing sea water are density, temperature and salinity. The relationship between these three is complex. The variation of temperature and salinity with depth is shown in Figure 2.6. In general, the temperature of the ocean decreases with depth. At the top of the ocean is the surface skin which may be only a few centimetres in depth. The main structure is defined as the mixed surface layer which stretches from the surface to tens of metres in depth. This is churned up by surface winds. Below this (about 100 m) there is a layer referred to as the **pycnocline**. This layer exhibits large changes in both temperature and density. The temperature decrease with depth in the pycnocline is called the **thermocline**. This exhibits seasonal changes and is well developed at low latitudes. The salinity increase with increasing depth is known as the **halocline** (Duxbury and Duxbury, 1989). The halocline is far more variable than the thermocline and can decrease with increasing depth at low latitudes (Thurman, 1993). Below the pycnocline the ocean becomes virtually isothermal with temperatures around 5°C. Salinity also becomes relatively stable. The structure of the ocean can be summarized as a relatively thin surface separated from the deep water by a thermocline. This temperature structure occurs because the water is heated from above. As the warm water is less dense and lies above the cold water this inhibits vertical motions and isolates the deep water. The more the surface water is heated and/or freshened by precipitation, the more difficult vertical motions become. This thermal stratification of water breaks down in the Arctic and Antarctic seas and is replaced by a single cold layer (see Fig. 2.6). It is at locations like these that vertical motions may take place.

MACRO-OCEANIC CIRCULATION: THE WINDS AND CURRENTS

Unlike the wind which is defined by the direction it has come from, an ocean current is defined by the direction it is flowing to. An eastward current in the ocean flows to the east in contrast to an easterly wind which blows to the west.

Surface circulation
The primary cause of surface currents are the prevailing surface winds. This was noted as early as the eighteenth century (Duxbury and Duxbury, 1989). Thermohaline circulations, that is circulations caused by solar heating, latent heat transference and salt movement (i.e. temperature and salinity) do occur, but only in certain areas. The coriolis force acts on the oceans just as it does on the atmosphere, deflecting the moving water. However, because water moves more slowly than air it takes longer for water to move the same distance as the wind. Thus the coriolis force has longer to act on the water and so can deflect the water to a larger extent than the overlying air. Figure 2.7 shows the main surface ocean currents of the world.

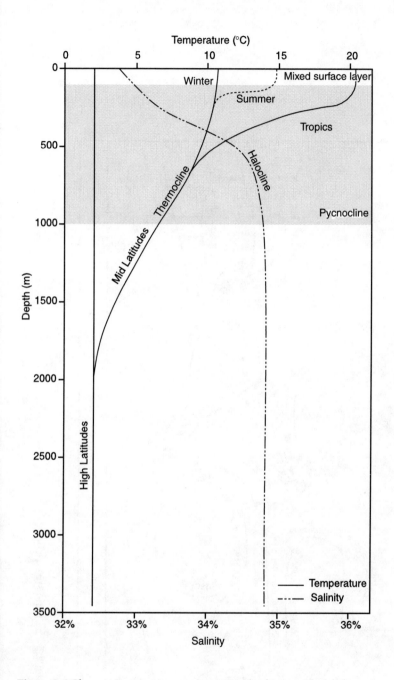

Figure 2.6 Changes in oceanic temperature and salinity with depth
Source: After Harries (1990)

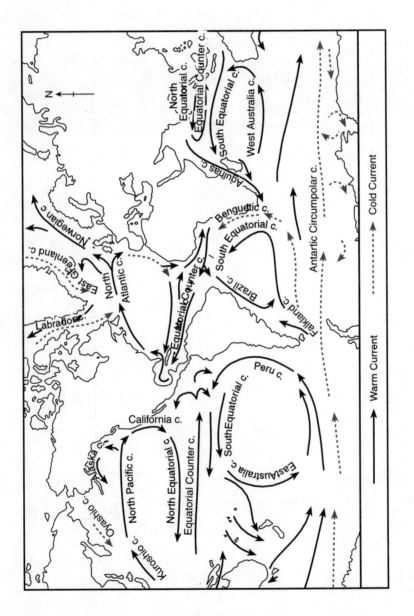

Figure 2.7 The main surface ocean currents of the world

Warm Current

Cold Current

The most prominent feature of this surface circulation are the large anti-cyclonic flows at approximately 30° latitude in all ocean basins except the northern Indian Ocean. They are called subtropical gyres. There tends to be subtropical convergence which leads to a piling up of water in the centre of these gyres perhaps up to 2m higher at the centre than at the edge. In gyres the coriolis force pushes water into the centre by means of Ekman transport. At the surface, the wind drives the surface water but due to low friction the layer of water below it will move more slowly than the surface layer. Consequently the coriolis force will have longer to act and the layer will be deflected more to the right in the northern hemisphere, and to the left in the southern hemisphere. Increasingly deeper layers will move more slowly and be deflected more creating a spiral current. This is called the **Ekman spiral** after V. Walfrid Ekman who first developed a mathematical treatment of the phenomenon (Walker, 1991). The depth of the spiral reaches to 100 – 150 m below the sea surface. At the bottom of the spiral the direction of the water flow will be at 180° to the surface current. Within the gyres the action of the coriolis force is then balanced by a gravitational force down the hillslope. If the two forces balance then the theoretical flow is parallel to the contours of the hill formed by the gyre.

The other main surface features are listed by Wells (1986). The most intense currents occur in the equatorial zone and along the east coasts of Japan (the Kuoshio current), North America (the Florida current and the Gulf stream) and South Africa (Agulhas current) with velocities of $1ms^{-1}$. In contrast, mid-ocean currents are weaker. In response to the surface winds there is an east–west nature to the currents in the equatorial Pacific and Atlantic. Also in the Southern Ocean an eastward current dominates. Most current boundaries, or frontal regions, are aligned in an east–west orientation. There are small anti-cyclonic circulations in the north-western Atlantic, the North Pacific near the Aleutian Islands and the Weddell Sea. In the northern Indian Ocean the semi-annual reversal in the surface winds due to the Monsoon results in an associated reversal of the surface ocean currents. The Monsoon is one of the prime examples of large scale air–sea interaction.

Deep water circulation

In winter at both poles sea-ice is formed; this is particularly so off Antarctica and the Weddell Sea. As sea-ice forms it expels the salt and the ocean immediately below the sea-ice becomes extremely saline. The sea water is also cold, and increasing cold with high salt content allows the sea water to sink downwards and then spread along the ocean floor. This explanation is rather simplistic as the Earth's rotation would actually stop the column of cold water from sinking and spreading. Send *et al.* (1996) explain four different ways in which the cold water may sink and spread. Essentially, however, the vertical sinking of water is due to changes in the salinity, density and the geometry of the ocean basin. In the Weddell Sea the sinking water has a salinity of 34.6

per cent and its temperature is $-1.9\,^{\circ}C$, nearly at the freezing point for water of this salinity. This water forms the Antarctic bottom water which has some of the highest densities (Thurman, 1993). On the ocean floor it is heated by geothermal energy. Arctic bottom water from the Arctic Ocean is usually confined and isolated from other waters. However, the **North Atlantic deep water (NADW)** generated south of Greenland is a substantial flow and mingles with water from the Gulf stream. Antarctic bottom water is free to spread and meets the NADW at 40° N. These deep cold waters must return to the upper ocean eventually. Sinking of deep water occurs very quickly but returning through the thermocline is very slow. This thermohaline circulation, shown in Figure 2.8, takes many years, approximately 1000 years for the ocean as a whole and this is called the **oceanic conveyor belt** (Broecker, 1987).

Currently the part of the conveyor belt nearer to the surface is warm and moving northwards. The deeper part is colder and moving south. This conveyor belt means that northern Europe is kept unusually warm. Paris lies 10 degrees of latitude farther north than New York and yet their mean temperatures are similar. In winter the shallow waters have been warmed on their journey up from the equator and tropics reaching the northern limit of the belt in Iceland. Here the air has come across from frozen Canada and there is a transfer of heat from sea to air, 25 per cent of that received by the North Atlantic from the Sun (Broecker and Denton, 1990). The consequence of this is that northern Europe is warmed and the sea is cooled, becomes dense and sinks to feed the lower part of the conveyor belt. The ocean acts as a pump taking heat from low latitudes and transferring it to high latitudes. The cold returning water goes round Africa and up into the Southern Indian Ocean and Pacific Ocean. In the North Pacific the conveyor belt operates the opposite way. This is a feature of our present-day climate. Why the oceans operate like this is not completely understood and there is increasing geological evidence to suggest that the conveyor belt does not always work in this way (Rahmstorf, 1997). Changes in the operation of the conveyor belt appear to have profound consequences for the climate which will be discussed in the next chapter. The oceanic conveyor belt also plays a part in maintaining the global water balance. There is an imbalance between oceans because the rate of evaporation from the Atlantic is higher than the Pacific. Thus water is transferred from the Atlantic via the atmospheric circulation to be precipitated over the Pacific. This leaves the Atlantic with a greater salinity, so there must be a flow of salt via the conveyor belt to the Pacific to maintain a balance. This is another example of the close coupling between the atmosphere and ocean.

Ocean–atmosphere interactions

The oceans form the surface boundary conditions for 70 per cent of the atmosphere. This leads to a number of ocean and atmosphere interactions. One of the most well known is the **El Niño Southern Oscillation** or ENSO

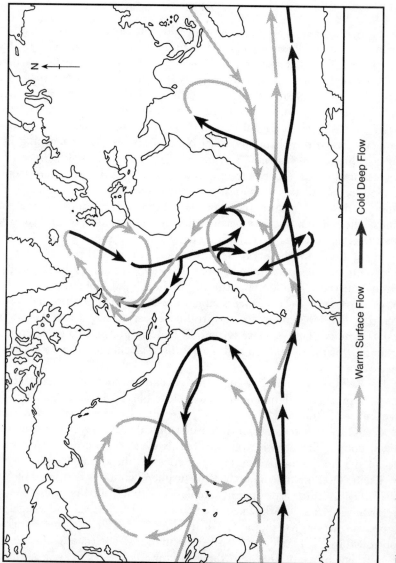

Figure 2.8 The deep circulation of the ocean termed the oceanic conveyor belt
Source: After Duxbury and Duxbury (1989)

Warm Surface Flow ⟶   Cold Deep Flow ⟶

for short. It is an interannual change resulting from internal interactions within the Earth's ocean–atmosphere system. It is not as a result of external forcings. The ENSO is an extremely important phenomenon which may have a global significance for the climate system.

EL NIÑO SOUTHERN OSCILLATION

For many centuries Peruvian fisherman had noted that the ocean waters in which they fished occasionally experienced unusually high temperatures (Philander, 1998). The fish and birds disappeared and the usually dry desert land became fertile as rains fell. These years were called the 'years of abundance' (Philander, 1990). It did not occur regularly and consecutive events could be anything from 2 to 11 years apart, but tended to be 2 to 5 years apart. The warmer temperatures started around Christmas time and the Peruvians called the phenomenon **El Niño**, a Spanish word meaning Little Boy or Christchild. It is now recognized that anomalously high sea surface temperatures occur about twice a decade in the central and eastern tropical Pacific Ocean and that these last for several months. At the same time an atmospheric event occurs and there is a notable weakening of the trade winds. The El Niño does not occur regularly and is said to be phased locked to the seasonal cycle, that is, it always happens in the same season. The maximum amplitude for the El Niño tends to occur around Christmas time. This statement is a little misleading because changes occur across the Pacific Basin throughout the year. The opposite of El Niño is **La Niña** and is a time of colder than average sea surface temperatures. There is not a normal state as such. The El Niño and La Niña constitute an aperiodic cycle and this is a natural variability of the climate system. El Niño is a term used to describe the ocean part of the oscillation and the Southern Oscillation is used for atmospheric component.

The ENSO is a climatic change which occurs without any external forcing and on a short enough time-scale to be able to observe both states. It provides an excellent example of ocean–atmosphere interactions. To fully understand the ENSO we have to consider the entire tropical Pacific Ocean. As Philander (1998) points out normal conditions hardly ever occur in the central Pacific Ocean. To aid the discussion, however, consider the average conditions over the Pacific Ocean basin in northern hemisphere winter as described by Wells (1986). The surface atmospheric circulation is controlled by a low pressure area which occurs over the maritime continent of Indonesia and a high pressure area over the North and South Pacific. As described earlier the flow is controlled by the Hadley and Walker circulations. The trade winds flow out of the high pressure area into the two convergence zones over the Pacific Oceans which are the ITCZ and SPCZ shown in Figure 2.9. The two areas of convergence are always found over the warmest surface waters (Philander, 1990). The tropical eastern Pacific between the two zones of convergence is an area of low precipitation and low sea surface temperatures. These cold sea

Normal conditions

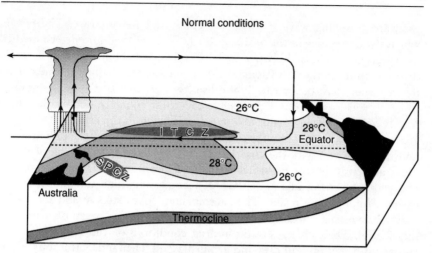

Figure 2.9 Typical atmospheric and oceanic conditions over the Pacific Ocean

surface temperatures are maintained by the strong south easterly trade winds. These winds produce the surface ocean circulation and effectively drive the warm surface water to the west and allow a pronounced upwelling of cold waters, by means of Ekman transport, along the equator and South American coast. A distinct west to east temperature gradient exists with cold water in the east, off the coast of Peru and warm water in the west.

It was noted in 1932 by Walker and Bliss that the atmospheric circulation pattern was not always maintained, that there were swings in it. The low pressure over Indonesia fills and the South Pacific high weakens. The normal Hadley–Walker circulation weakens and the south-east trades are replaced by westerlies. This is the **Southern Oscillation**. The difference in pressure between the Indonesian low and the Pacific high is known as the Southern Oscillation Index (SOI). When the SOI becomes low the main convergence zones also become displaced. The ITCZ moves southwards by 5 to 10 degrees and the SPCZ moves eastwards (Wells, 1986). This results in a change of the distribution of precipitation. The heaviest rainfall is displaced from Indonesia and the western Pacific to the Central Pacific and islands. In contrast, Indonesia and Australia suffer drought conditions. The slackening of the south-east trades causes the west–east temperature gradient in the sea surface temperatures to be lost. Harries (1990) describes this as the head of warm water built up in the west being released and sloshing back east. The reduced sea surface circulation restricts the amount of upwelling cold water. So warm water occurs off the South American coast and the El Niño occurs as shown in Figure 2.10a. The reality is slightly more complex than this as the localised anomaly in the south-east trades in the west central Pacific has to propagate through the entire Pacific Basin. This is achieved by waves that are generated internally in the upper ocean and move both west and east. The eastward travelling wave is called the equatorial **Kelvin wave**. The rate at which the anomalies in the SOI are spread east are therefore determined by the speed at which the Kelvin wave travels. It takes about 2 to 3 months for the wave to reach the east Pacific coast. Therefore the changes that occur off the Peruvian coast occur 1 to 2 months after the onset of the wind anomaly (Bigg, 1996). Rossby waves propagate the wind anomaly westwards, but it travels half the speed of the Kelvin wave. Therefore the effects at the west coast take two to four months to arrive. The interaction of the Kelvin waves and Rossby waves with the coastline eventually leads to the decay of the El Niño as waves are set up which leads to a restoration of the thermocline (Bigg, 1996). During a La Niña the SOI becomes very high and intense south-easterly trade winds result. The thermocline slopes steeply and far colder than average sea surface temperatures occur off the coast of Peru (*see* Fig. 2.10b). These changes result in drier conditions on the South American coast and heavy rainfall over India, Burma and Thailand. Variations in the position of warm and cold water can also lead to changes in the position of storm systems and the sub-tropical jet stream (Duxbury and Duxbury, 1989). One vexing question is what initiates the cycle? Is it changes in the ocean state or changes in the atmospheric state that cause an El Niño? As Philander (1998) points out, the atmosphere is quick to respond to changes while the ocean adjusts slowly with memories of the previous oceanic surface state propagating downwards through the thermocline. It would appear that the oceanic conditions are the stronger component.

The El Niño is no longer considered to bring good times but is perceived as a disaster. The increased precipitation now brings flooding and severe storms instead of bountiful harvests. This has more to do with human development of the region than changes in the El Niño (Philander, 1998). What has also been recognized is that the effects of the El Niño are felt far beyond Peru and the South American coasts. **Teleconnections** is the term used for how changes in weather patterns in one part of the globe can result in changes in other parts of the world. ENSO teleconnections have been well studied (Moron and Ward, 1998). ENSO has been shown to affect the subtropical Pacific basin and eastern and southern Africa. The most clear link is that between ENSO and the Indian monsoon. Typically during El Niño years the monsoon rains are weakened and there is less precipitation. There are various claims for the effects of the El Niño on the climate of northern Africa and Europe. Although climate anomalies can be shown to occur at these locations it is thought that the ENSO influence is small (Moron and Ward, 1998).

NORTH ATLANTIC OSCILLATION

The ENSO may be the most well-known oscillation but a similar one occurs in the Atlantic ocean. **The North Atlantic Oscillation (NAO)** is measured by an index which compares the pressure in Iceland and in the Azores. A high index is associated with low pressure in Iceland and high pressure off

[a]

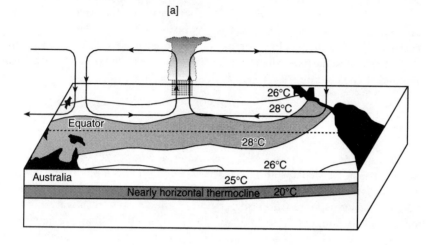

[b]

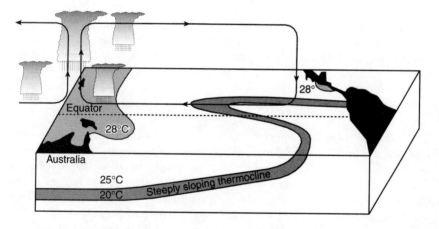

Figure 2.10 (a) El Niño conditions (b) La Niña conditions

Portugal in the Azores (Wilby *et al.*, 1997). While there are variations in the NAO throughout the year, the effect is greatest in winter (Hurrel, 1995). This brings warm strong westerly winds onto the European continent and in winter can result in gales. The 1930s was a time of high indexes and consequently strong surface westerlies. The warm winds meant that during this time Europe experienced a run of mild winters (Hurrel, 1995). When the index is low the pressure situation is reversed; this leads to winds from the north and the continent leading to cold European winters. This situation persisted from the 1940s until the early 1970s. The NAO is far more difficult to predict and appears to be chaotic, which would seem to suggest that the atmosphere is the controlling factor in the NAO. This is in contrast to ENSO where the oceans appear to be in control. However, as well as short-term variations there are longer decadal variations which may owe something to the ocean's long-term memory of the climate (Morton, 1998).

The cryosphere

The cryosphere encompasses all frozen water, including seasonal snowfalls, glaciers, the polar ice caps and permafrost. Once seen as a passive responder to the climate, it is now believed the cryosphere plays an active role in climate change. The cryosphere stores 80 per cent of the world's fresh water, but its importance to climate is in its high albedo for solar radiation and its low thermal conductivity (Peixoto and Oort, 1992). Also seasonal time-scales changes in the cryosphere accompany large-scale atmospheric circulation anomalies (Carleton, 1984). The cryosphere is split into two parts, seasonal and permanent. The seasonal ice cover (snowfall and sea-ice) responds dramatically to seasonal changes in the radiation budget. The permanent land ice sheets such as glaciers can show seasonal changes but the large continental ice sheets do not vary rapidly enough for seasonal changes to affect them. They play an important role in climate change over long time periods, up to tens of thousands of years, such as glacial and interglacial periods.

PERMANENT ICE COVER

Land ice sheets include the terrestrial ice sheets, such as the Arctic, mountain glaciers and the East Antarctic ice sheet, and also the marine ice sheet of West Antarctica. The difference is important because the marine ice sheet of the West Antarctic may respond more strongly to changes in the climate than the terrestrial ice sheets (Peixoto and Oort, 1992). These land ice sheets exert a strong control over the sea level and because they provide 10 per cent of the land surface are important for their interactions with the atmosphere. While the oceans store 94 per cent of the Earth's water, land ice stores only about 1.5 per cent (van der Veen, 1992). Table 2.3 shows the amount of water locked up in the major ice sheets and the effect it would have on sea level if it were all to melt. Glaciers and ice sheets are constantly growing and decaying. They increase in size by snowfall and decay either by ice bergs breaking off (**calving**) at the margins or melting. In Antarctica the low surface air temperatures prevent surface melting but ablation (melting) does occur at the bottom of the ice sheets (Peixoto and Oort, 1992). A balance between growth and decay is rarely achieved because the boundary conditions continuously change (van der Veen, 1992). The response times of the polar ice sheets to climate change is large, from 1 000 to 10 000 years for the Greenland ice sheet to 200 000 years for the east Antarctic ice sheets. In contrast, glaciers in mountainous areas respond rapidly to any changes in the radiative balance. While they only affect the local climate the instantaneous response time of glaciers make them a good indicator of climate change. It is difficult to attribute a cause to an observed change in an ice sheet because of the wide range of response times. Also feedback mechanisms may strongly amplify a glacier's observed response (van der Veen, 1992).

Table 2.3 Characteristics of terrestrial ice sheets

	Antarctica (grounded ice)	Greenland	Glaciers
Area (10^6 km^2)	11.97	1.68	0.55
Volume (10^6 km^3 ice)	29.33	2.95	0.11
Mean thickness (m)	2488	1575	200
Mean elevation (m)	2000	2080	-
Equivalent sea level (m)	65	7	0.35
Mass turnover time (yr)	15000	5 000	50 – 1000

Source: Houghton *et al.* (1990).

Seasonal Ice Cover

SNOW COVER

Seasonal snow cover is perhaps the most highly variable, both spatially and temporally, of all surface conditions. Snow cover has the highest albedo of any natural surface and as a consequence reflects a large portion of the incoming solar radiation which otherwise would go to heat the surface. Consequently large-scale anomalies in the amount of snow can seriously affect the surface energy budget. Changes in the surface heating will result in changes in the heating of the atmosphere which in turn can affect the climate (Cohen, 1994). It is also thought to restrict the availability of water and thus affects the water as well as the heat budget (Kane, 1992). Cohen (1994) reviews the affect of seasonal snow cover on the local temperature and identifies four ways in which the temperature is affected.

1. By increasing the surface albedo by 30–50 per cent freshly fallen snow reduces the amount of solar radiation absorbed at the surface leading to lower daytime temperatures.
2. Snow increases the amount of infra-red radiation lost near the ground resulting in a lower temperature minimum near the surface.

3. As snow is a good insulator heat is not released from the ground to the overlying atmosphere.
4. Snow requires energy (latent heat) in order to melt. Thus heat is taken from the surrounding ground and atmosphere, so cooling them, to melt the snow.

The combined effect of these on the surface energy budget is to produce a large loss of energy over a snow field. This is predominantly due to the increase in the surface albedo. The atmosphere overlying a snow field therefore becomes cold and cannot hold very much moisture. It is perhaps easy to comprehend how these changes resulting from seasonal snow cover affects the local climate. There are several studies, however, which suggest that snow cover may induce more widespread changes. For example, extensive snow cover is correlated with delays in the onset of spring warming, and the intensification of autumn/winter cooling. There is also evidence to suggest that snow cover might also affect the development of depression systems over the east coast of North America. The depression systems develop farther south and occlude more quickly with extensive snow cover. Such large-scale changes rely on 'atmospheric teleconnections' whereby changes in the atmospheric conditions in a local area are propagated to far-away areas through the internal dynamics of the atmosphere (Cohen, 1994).

SEA-ICE

Sea-ice forms in polar regions due to the extreme low temperatures. It forms from freezing sea water and accumulated snow and is distinct from ice bergs which form from calving. In some areas, such as the Arctic near Siberia, sea-ice occurs all year round. Whereas in other regions around the Arctic and Antarctica sea-ice is a seasonal phenomena; growing in winter and melting in summer. The growth of the sea-ice around Antarctica in winter significantly increases the size of the southern polar ice cap. There are distinct differences in the formation of sea-ice between the two polar regions mostly due to the different geographical make-up of the areas. Antarctic sea-ice grows slowly over seven months but retreats quickly (Massom, 1991). In contrast Arctic sea-ice grows quickly but decays slowly. Sea-ice begins as small ice crystals which form a thin slush layer on the ocean surface. This thin ice sheet is broken up by wave action to form pancake ice. These pancakes are further carried by the wind and waves colliding into each other. As the freezing and collisions continue the pancakes form ice floes. As the sea-ice forms salt is expelled. The ice is formed from nearly fresh water and thus the water directly below the sea-ice increases in salinity. Therefore the formation of sea-ice at polar regions is an important first part of the deep oceanic conveyor belt. If the ice melts, releasing fresh water, convective activity can stop. Salt water is trapped in the sea-ice in brine pockets. However, as the ice ages these are flushed out and the sea-ice becomes increasingly fresh.

Sea-ice forms by freezing the layer of water closest to the sea-ice bottom surface. Therefore the thickness of sea-ice is limited by the temperature at the bottom. The temperature at the surface is immaterial (Niiler, 1992). In its first year sea-ice grows to a thickness of about 2 m. The mean thickness of Arctic sea-ice is 3.0 to 4.0 m and is much thicker than Antarctic sea-ice which is usually only 0.3 to 1.0 m. This is because in the Antarctic the ocean is unimpeded and there is a large oceanic heat flux. In fact, in the Antarctic sea-ice growth is far from orderly due to the stormy southern ocean and the ice cover is principally formed from small ice particles which agglomerate to form frazil ice (Hibler and Flato, 1992). Strong winds and ocean currents can create large ridges in the ice underwater. This increases the drag from the ocean on the sea-ice and may eventually cause long rectangular channels to form called **leads**. The leads are usually tens of metres wide and thousands of metres long. The heat loss from the ocean due to these breaks in the ice can be large. The wind can increase the size of the leads creating **polynyas** which are big round expanses of open water typically tens or hundreds of kilometres in diameter (Hanna, 1999). Thin ice grows much faster than thick ice. This is because thin ice does not insulate the ocean from heat loss as well as thick ice. In the same way leads enhance the formation of sea-ice. Ridging leads to thin ice being piled up on thick ice forming triangular features which can survive the summer melt (Hibler and Flato, 1992).

The effect of sea-ice on the oceanic energy budget is similar to that of snow cover on the land energy budget. Sea-ice has a low thermal conductivity so it insulates the sea below it, allowing less heat to escape into the atmosphere above. It also affects the moisture and momentum transfers between the atmosphere and ocean (Massom, 1991). The main effect, however, is the ice-albedo. The sea-ice is a highly reflecting surface, therefore solar radiation is reflected away, cooling the surface. At present there is a great deal of interest in the monitoring of Antarctic sea-ice as it may be particularly sensitive to climate changes resulting from global warming (Hanna, 1999). The interactions between sea-ice and the rest of the climate system are complex but several links have been noted. The extent of sea-ice in Antarctica seems to affect cyclogensis. There are also links to ENSO with the strongest being that ENSO leads the sea-ice by 6 to 12 months (Hanna, 1999). The real importance of sea-ice to the climate system is through the positive ice-albedo feedback. If the sea-ice should melt, then the dark sea water could absorb more solar radiation increasing the temperature and thus melt more sea-ice. The effect is made more complex due to the presence of clouds and their significant impact on the radiation budget (Walker, 1998). There is good reason to believe that cloudiness should be investigated in conjunction with the cryosphere, as many surface–atmosphere feedbacks important to climate seem to be amplified at snow/ice boundaries which appear to be a preferred area for cloud formation (Carleton, 1984).

The biosphere

The **biosphere** includes plants on land and in the sea and all animals. The response time of the biosphere is variable and not well defined. This reflects not just a lack of knowledge, but also the diversity of the biosphere. The land surface covers approximately 30 per cent of the Earth and, while some may be ice covered, much of the surface will be covered with vegetation. Therefore vegetation provides the boundary layer to a substantial proportion of the Earth's atmosphere. The vegetation will therefore determine the surface energy budget, with characteristics such as the albedo and latent heat transfer being modulated by the type of vegetation. The atmospheric boundary layer, the lowest part of the atmosphere, interacts directly with the surface vegetation. The drag on the atmosphere will be affected by the height of the vegetation. For a long time such issues have mainly been of interest to micro-climatologists but it is increasingly important to the global climate debate. The biosphere not only directly affects the radiation budget through changes in albedo but also mass exchange, in particular water and carbon. Plants affect soil structure through leaf litter and also their root system. This in turn affects the interception of precipitation and the way in which water runs off and infiltrates the soil. Some of the most disastrous localized impacts of extensive deforestation have been the mudslides caused by the changes in run-off. The catchment area of the Yangtze River in China has been extensively deforested and heavy rains during 1998 resulted in several devastating floods. The El Niño originally took the blame but the Chinese government has recently acknowledged that the more likely culprit is the removal of forest from the area.

The biosphere is not only a result of the climate, but may well play a vital modulating role. This is through the carbon dioxide cycle that was described in the section on the hydrosphere. The importance of the biosphere to this cycle cannot be stressed too strongly. All aspects of the biosphere regulate the cycle. For example, young trees take up more carbon dioxide than mature trees. Therefore changes in the biosphere, whether natural or human, may result in climatic change. Furthermore the sensitivity of the biosphere to the atmospheric climate means that past climates can be studied through their effect on the biosphere.

The lithosphere

The **lithosphere** is the surface crust of the Earth, the mountains, rocks and ocean basins. While it does interact with the overlying atmosphere, by transfers of mass and energy, its very long response time means that except for geological time-scales, the lithosphere can be regarded as unchanging. On geological time-scales continental drift, uplift and subsidence may all play a role in the modulating the climate.

Interactions: feedbacks and sensitivity

The sub-systems of the climate are not always in equilibrium with each other or in internal equilibrium. The sub-systems also have feedback loops. Feedback is a term coined from electronics where a change in the original signal results in a change in output which then feeds back into the signal again. It is these feedbacks which make the climate system so complex as an original forcing to the system will be considerably altered by these feedback loops. Feedbacks can be positive or negative. Positive feedbacks are where the original signal is amplified. In negative feedbacks the signal is damped. Positive feedbacks are much more common in natural systems than negative feedbacks. A good example of a positive feedback in the climate system is the ice-albedo feedback. The distribution of snow and ice cover depends primarily on the near surface air temperature. If a perturbation occurs which decreases the temperature, more snow and ice will be able to form. Snow and ice are highly reflective and so increased snow cover will reflect more solar radiation leading to decreased near surface air temperatures. Similarly, as the temperature increases so the amount of snow and ice will decrease and the amount of solar radiation absorbed will increase and the temperature will rise. By contrast, one of the few negative feedbacks to be identified relates to the Stefan-Boltzmann Law. If the temperature of the atmosphere increases, then it will lose more radiation to space (by the Stefan-Boltzmann law). This will reduce the temperature and ameliorate the original change (Peixoto and Oort, 1992).

Sometimes it is not always clear which feedback dominates. A good example of this are clouds. Clouds reflect and scatter incoming solar radiation. As much as 80 per cent of incoming solar radiation may be reflected back into space by clouds (Coley and Jonas, 1999). This albedo effect of clouds leads to a cooling of the Earth, an apparently negative feedback. Clouds, however, are composed of water which is a very good absorber of long-wave radiation, therefore clouds have a greenhouse effect. Clouds warm the Earth by trapping the outgoing long-wave radiation which is a positive feedback. On a global annual average clouds act to cool the Earth. Generally low clouds act to cool the Earth while high thin cirrus clouds act to warm the Earth (Coley and Jonas, 1999). The large control that clouds have on the radiation balance means that clouds are an important area of study for climatologists. It is fairly easy to see that a change in cloud amount will affect the Earth's radiation budget and clouds may well change in response to global warming. This aspect of cloudiness will be discussed further in Chapter 6.

Due to feedback even for a forcing, which could be measured quite accurately, it is very difficult to predict how the climate system will respond. Through the various feedback mechanisms the climate might be quite stable to large forcings or it might be extremely unstable and change in response to very small forcings. Climatologists therefore talk in terms of the sensitivity of the climate to a forcing. **Sensitivity** is (as defined by Houghton *et al.*, 1990) the change in surface air temperature in response to a radiative forcing.

Equilibrium and stability

If the solar output was to suddenly double, the various components of the climate system would all start to respond in line with their response times. Eventually a new energy balance would be reached and the climate would be in equilibrium. Chapter 1 introduced equation (1.9) to calculate the temperature at the top of the atmosphere; the energy balance equation.

$$T^4 = So(1-a)/4\varepsilon\sigma \tag{2.2}$$

That equation was solved to find T by using present-day values of S_0 and a. If the solar output doubled you would obviously get a new temperature. This would be the new **equilibrium** point, the new balance between incoming and outgoing energy, when the climate system has responded fully to the forcing. As reaching a new equilibrium is not instantaneous, the climate is said to be in transition. This **transient** climate will represent part of the equilibrium response and the temperature at any time is said to be the realized temperature change. In reality, the nature of the forcing is unlikely to occur instantaneously. Much more likely is a gradual change over time in solar output. Similarly with global warming, as will be shown in Chapter 5, there has been a gradual increase in anthropogenic greenhouse gases. The forcing is gradually changing and the climate's response to that change is also gradual. It may take the oceans hundreds of years to respond to a change in forcings. Therefore even if all the greenhouse gas emissions were to stop tomorrow, the climate would still be committed to a change. This would be in line with the degree of change that has already occurred in the forcing.

Furthermore, the climate system is full of occurrences of feedback. There is even one in the simple equation above, that is albedo and the ice-albedo feedback. Albedo is not a constant but is temperature dependent as described in the previous section. Therefore in the above equation albedo can be rewritten as a function of mean global surface temperature (T_s). When the mean surface temperature is high enough there will be no snow or ice and the albedo will be independent of temperature. The albedo can then be assigned a relatively low but constant value. Similarly, when the planet is cold enough it will become covered with snow and ice and again the albedo will be independent of the surface temperature. This time the albedo will take a high constant value. In between these two extremes the planet will be partially ice covered, with the amount of ice cover, and thus albedo, depending on T_s. Then a simple empirically based equation can be derived between albedo and T_s. A simple equation for outgoing radiation and T_s can also be obtained from observations. Now both the outgoing and incoming radiation can be calculated for various values of T_s (see Kiehl, 1992; note in this book the value for the solar constant used is 1368 Wm^{-2}). The results from these are plotted on a graph in Figure 2.11. The points on the graph where the two lines cross are equilibrium positions when the outgoing radiation balances the incoming radiation. There are three equilibrium points: (1) completely glaciated;

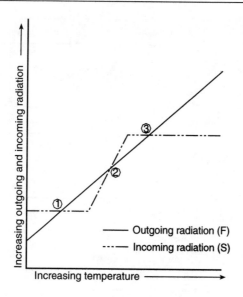

Figure 2.11 Solutions to the incoming solar radiation (S) and outgoing long-wave radiation (F) plotted against temperature revealing three equilibria

(2) partial glaciation; and (3) ice free. Therefore this simple model indicates that the climate system may have more than one equilibrium position. That is it may be possible for the climate system to exist in any one of these states.

But are all these states stable; that is will they persist for long periods of time? To answer this imagine a cone placed on its side as in Figure 2.12b. If you push it then it may roll around but it will stay on its side. This position is said to be neutrally stable. If you place it on its base (Figure 2.12a) then, although a very hard push may knock it over, the cone reacts to any force by returning to its original position on its base. If you balance a cone on its pointed end (Figure 2.12c) then even a slight force will cause it to fall over on to its side. All these positions are equilibrium points but some are more

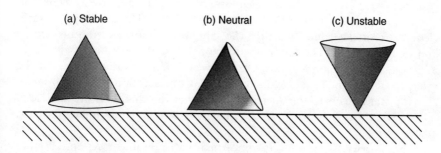

Figure 2.12 Examples of (a) stable, (b) neutrally stable and (c) unstable conditions

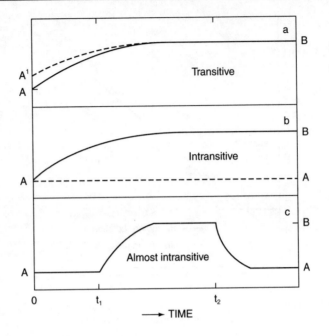

Figure 2.13 Concepts of (a) transitive, (b) intransitive and (c) almost intransitive
climate system
Source: Peixoto and Oort (1992)

stable than others. In the case of the cone, it is most stable on its side and least
stable on its tip. Figure 2.13 illustrates possible behaviours of the climate system. If there are two different starting points (initial conditions) but the system always evolves to one state, B, then the system is said to be transitive. Any
perturbations will eventually lead back to B, so B is a stable equilibrium
point. If, however, the system can evolve to different states from the same initial condition, then the system is called intransitive. An almost intransitive
system from initial conditions alternates between different resultant states.
Many non-linear systems do this and many of the equations that govern the
climate system are non-linear. Glacial/interglacial periods maybe a sign of an
almost intransitive climate (Peixoto and Oort, 1992). There is, as yet, insuffi-
cient data to determine whether the climate is transitive, intransitive or
almost intransitive. The latter would be particularly difficult to model. In
addition, if the climate is almost intransitive this almost-intransivity could be
interpreted by humans as a climate change. Consideration of these different
climatic states leads to the question of whether the climate is predictable at
all and what role chaos theory may play in determining the climate (Lorenz,
1968).

Chaos theory is particularly concerned with non-linear systems. Rather
than define a **non-linear equation** it is easier to describe a **linear equation**. A
linear equation is one which allows you to draw a straight line. There is a

one-to-one relationship between the X and Y values on a graph. Trying to draw the curve from a non-linear equation would result in a much more complex pattern. A simple system that is ruled by a non-linear deterministic equation is the pendulum. You can start the pendulum off from an exact known position at an exact time and from the equations predict, or determine, where it will be in 3 seconds, 4 minutes, etc. As stated earlier, the equations governing the climate system form a non-linear deterministic system and it was thought that these systems were exactly predictable. However, the branch of mathematics called chaos theory is based on the fact that such systems are not necessarily predictable. It turns out that even the simple example, of the pendulum, is extremely sensitive to the initial conditions (Baker and Gollub, 1990). If you set a pendulum swinging from an initial state x and then from x + e, where e is a very small quantity, their dynamical states will diverge quickly, the separation increasing exponentially. Very small differences in initial conditions leads to large differences in the final result. 'Prediction becomes impossible' (Poincaré, 1913). From this statement Baker and Gollub (1990) conclude that a chaotic system must resemble that of a stochastic system. That is a system that is subject to random forces. In a chaotic system, however, the irregularities are internal features of the system, not external.

As the equations that govern the climate system form a deterministic non-linear system, given the initial conditions, it should be possible to solve the equations for some time later. Determining the initial state of the climate, however, would be an impossibility. This leads to the uncomfortable conclusion that weather and climate may never be predictable. It has been suggested that even the movement of a butterfly's wing will cause major changes in the weather end system. Some parts of the climate system may well be chaotic on time-scales of a century to a millennium. Weather systems, such as cyclonic activity, in both the atmosphere and the ocean are undoubtedly chaotic. In the atmosphere the large horizontal scale of these systems make it necessary to consider the whole atmospheric motions as chaotic. There are, however, other features which are stable such as the seasonal cycle in temperature distribution over the continents, monsoons and storm tracks. It is the stability of these features which lead many climatologists to conclude that the response of the climate system to any external forcing will be stable (Cubasch and Cess, 1990). The stable behaviour of the climate system means that we can look for changes by taking time averages. In order to confidently attribute any changes to an external forcing, however, the time averages have to be long compared to any chaotic behaviour.

Global warming

The Earth's atmosphere acts just like a blanket. On a cold winter's night you feel nice and warm but put your hand outside on top of the blanket and it's a lot colder. The Earth's surface remains far warmer than the top of the atmosphere because the atmosphere traps the heat. Water, particularly water vapour, carbon dioxide and other trace gases in the atmosphere are all very effective absorbers of infra-red radiation. They absorb the long-wave radiation coming up from the surface of the Earth. When they re-emit the radiation, they do so in all directions; some will be lost to space but some will go back down to be reabsorbed by the Earth's surface, increasing the surface temperature of the Earth. Water vapour and carbon dioxide are referred to as greenhouse gases. If there is a greenhouse gas in the atmosphere, this effect will occur.

The problem is that human activity is increasing the amount of greenhouse gases in the atmosphere. This increases the amount of long-wave radiation that can be absorbed and therefore the amount that can be re-emitted back to warm up the Earth. Most of this absorption and re-emission is going on in the troposphere, the lower atmosphere. That is also where humans are increasing the amount of greenhouse gases. The Earth is losing less infra-red radiation, but it has to maintain the balance with the incoming radiation, and the only way to do this is to increase the radiating temperature. Both the troposphere and the surface warms. It's like throwing another blanket on the bed, you get warmer. The idea of a change in radiative forcing leading to a change in surface temperature comes from climate model results that show there is a linear relationship between the global-mean radiative forcing at the tropopause and the equilibrium global-mean surface temperature change (Shine *et al.*, 1994). But what happens at the top of the atmosphere? If there were no feedbacks within the system and nothing else changed, then the top of the atmosphere radiances would remain the same, and so would the planetary effective temperature. Indeed, under these circumstances it would be relatively easy for climate modellers to predict surface temperature changes. The climate system, however, is not like that and there are many other variables that might change. For example, the planetary albedo might increase due to increased cloud cover. This would reflect away more solar radiation so less solar radiation would play a part in the climate system and the planetary effective temperature would drop. The relationship between global mean radiative forcing (ΔQ) and the global mean surface temperature (ΔT) can be written mathematically as

$$\Delta T = \lambda \Delta Q \qquad (2.3)$$

where λ is given by

$$\lambda^{-1} = \Delta F/\Delta T - \Delta S/\Delta T \qquad (2.4)$$

λ is called the **climate sensitivity** parameter and is there to account for all the feedback processes, such as the cloud feedback described above, within the climate system. The whole temperature structure of the atmosphere is affected by global warming. The increase in temperature of the surface–troposphere system will cause the stratosphere to cool. It has to do this in order to maintain the radiation balance with the top of the atmosphere.

Much of the controversy that now surrounds the issue of global warming is not whether global warming will occur but just how sensitive the climate is to increases in carbon dioxide levels.

The idea of greenhouse warming is not a new one and neither is the argument (Mudge, 1997). Back in 1827 a French scientist Fourier had suggested the atmosphere was like a 'hot-house' in that it let through the 'light-rays' of the Sun but trapped the 'dark rays' from the Earth. In 1861 John Tyndall had suggested that, like water vapour, carbon dioxide gas could act as a good absorber of 'calorific rays'. In 1896 a Swedish scientist called Arrhenious calculated that a doubling of carbon dioxide would result in a 8°C rise in global temperature. By 1897 Chamberlain, an American geologist, had developed the carbon dioxide theory for driving climate changes (Mudge, 1997). By looking at the Earth's two nearest planetary neighbours, Venus and Mars, we can observe how carbon dioxide acts as a greenhouse gas and how the amount is important for determining the surface temperature. Also, there is ample geological evidence to suggest that the carbon dioxide content of the Earth's atmosphere has played an important role in determining the climate and perhaps the evolution of life itself.

References

Arrhenious, S. 1896: On the influence of Carbonic acid in the air upon the temperature of the ground. *Philosophical Magazine* **41**, 237–76.

Baker, G.L. and Gollub, J.P. 1990: *Chaotic dynamics: an introduction.* Cambridge: Cambridge University Press.

Barry, R.G. and Chorley, R.J. 1992: *Atmosphere, weather and climate.* Sixth edition. London: Routledge.

Bigg, G.R. 1996: *The oceans and atmosphere.* Cambridge: Cambridge University Press.

Bigg, G.R. 1997: Light in the atmosphere: part 1 - why the sky is blue. *Weather* **52**, 72–7.

Broecker, W.S. 1987: The biggest chill. *Natural History Magazine*, October, 74–82.

Broecker, W.S. and Denton, G.H. 1990: The role of ocean–atmosphere reorganisations in glacial cycles. *Quaternary Science Reviews* **9**, 305–41.

Carleton, A.M. 1984: Cloud–cryosphere interactions. In Henderson-Sellers, A. (ed.), *Satellite sensing of a cloudy atmosphere.* London: Taylor and Francis.

Cohen, J. 1994: Snow cover and climate. *Weather* **49**, 150–6.

Coley, P.F. and Jonas, P.R. 1999: Clouds and the Earth's radiation budget. *Weather* **54**, 66–70.

Cubasch, U. and Cess, R.D. 1990: Processing and modelling In Houghton, J.T., Jenkins, G.J. and Ephraums, J.J. (eds), *Climate change: the IPCC scientific assessment.* Cambridge: Cambridge University Press.

Drake, F. 1995: Stratospheric ozone depletion: an overview of the scientific debate. *Progress in Physical Geography* **19**, 1–16.

Duxbury, A.C. and Duxbury, A.B. 1989: *An introduction to the world's oceans.* Dubuque: Wm. C. Crown.

Hanna, E. 1999: Recent observations of Antarctic sea-ice. *Weather* 54, 71–86.

Harries, J.E. 1990: *Earthwatch: the climate from space.* Chichester: Ellis-Horwood.

Henderson-Sellers, A. and Robinson, P.J. 1986: *Contemporary climatology.* London: Longman Group.

Hibler W.D. and Flato, G.M. 1992: Sea ice models. In Trenberth, K.E. (ed.), *Climate system modeling.* Cambridge: Cambridge University Press.

Houghton, J.T., Jenkins, G.J. and Ephraums, J.J. (eds) 1990: *Climate change: the IPCC scientific assessment.* Cambridge: Cambridge University Press.

Hurrel, J.W. 1995: Decadal trends in the North Atlantic Oscillation and its relation to regional temperature and precipitation. *Science* 269, 676–9.

Kemp, D.D. 1990: *Global environmental issues: a climatological approach.* London: Routledge.

Kiehl, J.T. 1992: Atmospheric circulation models. In Trenberth, K.E. (ed.), *Climate system modeling.* Cambridge: Cambridge University Press.

Lockwood, J. 1979: *Causes of climate.* London: Arnold.

Lorenz, E.N. 1968: Climate determinism. *Meteorological Monographs* 8, 1–3.

Massom, R. 1991: Satellite remote sensing of polar regions. London: Belhaven Press.

McIlveen, R. 1992: *Fundamentals of weather and climate.* London: Chapman and Hall.

Moron, V. and Ward, M.N. 1998: ENSO teleconnections with climate variability in the European and African Sectors. *Weather* 53, 287–95.

Morton, O. 1998: The storm in the machine. *New Scientist* 157 (no. 2119), 22–7.

Mudge, F.B. 1997: The development of the 'greenhouse' theory of global change from Victorian times. *Weather* 52, 13–17.

Niiler P. P. 1992: The ocean circulation. In Trenberth, K. E. (ed.), *Climate System Modeling.* Cambridge: Cambridge University Press.

NRC 1994: *Solar influences on global change.* Board on global change commission on geosciences, environment and resources, National Research Council. Washington, DC: National Academy Press,

O'Hare, G. 1997a: The Indian Monsoon Part 1: The Wind System. *Geography* 82, 218–30.

O'Hare, G. 1997b: The Indian Monsoon Part 2: The Rains. *Geography* 82, 335–52.

Peixoto, J.P. and Oort, A.H. 1992: *Physics of climate.* New York: American Institute of Physics.

Philander, G. 1990: *El Niño, La Niña and the Southern Oscillation.* London: Academic Press.

Philander, G. 1998: Learning from El Niño. *Weather* 53, 270–4.

Poincaré, H. 1913: *The foundation of science: science and method.* English translation, 1946. Lancaster, PA: The Science Press.

Rahmstorf, S. 1997: Ice-cold in Paris. *New Scientist* 153 (no. 2068), 26–30.

Send, U., Font, J. and Mertens, C. 1996: Recent observations indicates convection's role in deep water circulation, *Eos* 77, 61.

Shine, K.P., Fouquart, Y., Ramaswamy, V., Solomon S., and Srinivasan J., 1994: Radiative forcing. In Houghton, J.T., Meira Filho, L.G., Bruce, J., Hoesung Lee, Callender, B.A., Haites, E., Harris, N., and Maskell, K., (eds), *Climate change 1994: radiative forcing of climate change and an evaluation of the IPCC IS92 emission scenarios.* Cambridge: Cambridge University Press.

Solomon, S. 1988: The mystery of the Antarctic ozone 'hole'. *Review of Geophysics* 26, 131–48.

Stanton, H.R. 1991: Ocean circulation and ocean–atmosphere exchanges. *Climatic Change* 18, 175–94.

Thurman, H.V. 1993: *Essentials of oceanography. 4th edition.* New York: Macmillan Press.

van der Veen, C.J. 1992: Land ice and climate. In Trenberth, K.E. (ed.), *Climate system modeling.* Cambridge: Cambridge University Press.

Walker, G. 1998: On thin ice. *New Scientist* 159 (no. 2153), 32–7.

Walker, G.T. and Bliss, E.W. 1932: World weather V. *Memoirs of the Royal Meteorological Society* 4, 53–84.

Walker, J.M. 1991: Farthest north, dead water and the Ekman spiral. Part 2: invisible waves and a new direction in current theory. *Weather* 46, 158–64.

Wells, N. 1986: *The atmosphere and ocean: a physical introduction.* London: Taylor and Francis.

Wilby, R.L., O'Hare, G. and Barnsley, N. 1997: The North Atlantic Oscillation and British Isles climate variability, 1885–1996. *Weather* 52, 266–76.

|3|

Past greenhouses and icehouses

Introduction

This chapter considers the evolution of the Earth's atmosphere and its planetary neighbours. In doing so, examples of the natural greenhouse effect and its importance for evolution of life on Earth are considered. There are many who would argue that the Earth's climate and life are inextricably linked. After looking at the very early Earth the chapter will discuss the quaternary glacial cycles and the mechanisms which may explain the waxing and waning of the permanent polar ice sheets.

Evolution of the planetary atmosphere

The gases that make up a planetary atmosphere are important for the climate of the planet and its suitability for life. The gaseous composition of the atmosphere will depend on the planet's formation and subsequent history. It also depends on whether the planet is able to hold on to any atmosphere formed. A planet's ability to retain a gas is dependent on the gravitational pull and temperature of the planet, and the mass of the gas. To fully appreciate the Earth's climate we need to consider how the Earth and the other planets obtained their atmospheres and understand what has happened to them since the solar system began.

There have been many mechanisms suggested for the formation of the solar system. The most widely accepted theory is the planetesimal hypothesis (Greenberg, 1989). The starting point is a swirling mass of gas and dust: interstellar matter called the primordial solar nebula. The nebula is believed to have had the same composition as the outer regions of the Sun today and was rotating slowly. About 4.6×10^9 years ago compression of the interstellar matter and heating and cooling, caused by its own gravitational attraction, made the nebula contract and rotate faster and faster. The cloud began to fall in on itself due to this attraction, leaving behind a disk of cooling nebula. The central condensation became the Sun. From the cold rotating disk the planets were formed. First the gaseous nebula condensed into solid particles and then these solids stuck together in a process called **accretion**. The accreting particles all moved in the same direction so collisions tended to result in lumps sticking together rather than breaking each other apart. As the planets grew larger they swept up any smaller particles that stood in their path. The remaining solar nebula then eventually dissipated.

From their characteristics the planets are usually split into two main groups, the terrestrial (Earth-like) planets; Mercury, Mars, Earth, Venus and the gas giants; Jupiter, Saturn, Uranus and Neptune (*see* Table 3.1). As can be seen from Table 3.1, Pluto, the outermost planet in the solar system has very little in common with the rest of the planets. This has led to speculation that Pluto is a captured planet, from another solar system, rather than formed as part of our solar system. However, more recently it has been suggested that Pluto is a remnant of the accreting process. In the cold outer regions of the solar system the accreting particles consisted of large icy bodies. Most of these were swept up to form the gas giants and the remaining smaller ones form the Kuiper belt. Pluto represents the intermediate stage and is a surviving small planet that avoided being swept up by its larger neighbours. In between the terrestrial planets and the gas giants lies the asteroid belt. The asteroids are the remains of the accreting material that cannot form another planet because the gravitational attraction of Jupiter pulls them apart. There are four possibilities as to the origin of the planetary atmospheres. The Sun sends out a continuous stream of charged particles called the solar wind. In the early solar system it may have been much stronger, called a T-Tauri wind. One possibility is that the planets captured this solar wind to form an early atmosphere. Alternatively, the planets could have obtained their atmospheres from asteroids and comets. Both of these contain gases, or volatiles, which if they collided with a planet would be released forming an atmosphere. Another possibility is the capture and retention of the primordial solar nebula. Finally the rocky cores of the planet, just like comets and asteroids, would contain volatiles which could be outgassed to form an atmosphere. By looking at the composition of the planetary atmospheres it is possible to deduce the likely mode of formation.

In the universe the most abundant elements are hydrogen, then helium. Looking at the Sun we see the same thing, hydrogen and helium. The giant planets' atmospheres are also predominantly hydrogen and helium. Saturn and Jupiter grew large enough to retain and attract the solar nebula, as the solar nebula dissipated after the formation of the solar system. Uranus and Neptune are far smaller and consequently only retained a residual amount of the solar nebula surrounding them. The giant planets may have formed in much the same way as the Sun as sub-condensations on the disk. When we look about the universe we see many binary star systems. An old theory of

Table 3.1 Comparative table of planets

Planet	Distance from the Sun given Earth is 1 (1 astronomical unit)	Mass given Earth is 1	Radius given Earth is 1	Axial tilt (degrees)	Day length	Year length (years)	Planetary effective temperature (°C)	Surface temperature (°C)	Surface pressure (mb)	Significant atmospheric gases
Mercury	0.4	0.06	0.4	0	59 days	0.2		350 (day) −170 (night)	0	None
Venus	0.75	0.8	1	2	243 days (retrograde)	0.6	−38	480	90 000	CO_2
Earth	1	1	1	23.27	24 hours	1	−33	22	1000	N_2, O_2, H_2O, Ar, CO_2
Mars	1.5	0.11	0.5	24.46	24.5 hours	1.9	−53	−23	6	CO_2, Ar
Jupiter	5	318	11	3	10 hours	11.9	−173			H_2, He
Saturn	10	95	9	27	10 hours	29.5	−198			H_2, He
Uranus	20	15	4	98	11 hours	84.0	−223			H_2, He, CH_4
Neptune	30	17.2	4	29	16 hours	164.8	−233			H_2, He, CH_4
Pluto	40	2.1	0.5		6 days	248.4	−233 (?)			N_2(?)

planetary formation was that the planets were formed from the Sun's companion star. More recently it has been suggested that Jupiter may be a proto-companion star. Jupiter is a planet but it has a mass nearly three times that of the rest of the planets put together. The gravitational attraction of Jupiter is such that it can generate its own heat (Henderson-Sellers, 1983).

In contrast, the terrestrial planets are believed to have formed, from the accretion of smaller solid particles. After the terrestrial planets had formed, their gravitational fields may have been large enough to attract a small residual of solar nebula gases, that were left behind once the solar nebula dissipated. This is often referred to as the primordial or primary atmosphere. It would have been composed of hydrogen, helium, methane and ammonia and is referred to as a reducing atmosphere. A reducing atmosphere is one where free oxygen and highly oxidized gases cannot occur. It was once thought that such an atmosphere was a prerequisite for the evolution of life. There is no evidence, however, to suggest that such an atmosphere, if it existed, survived much beyond four thousand million years ago. Ammonia is easily photodissociated and, once removed, there is no obvious way to replenish it in the atmosphere. This primordial atmosphere, if it existed, must have been almost completely dissipated; perhaps swept away by the strong solar wind.

The secondary atmosphere, in contrast to the primary atmosphere, was rich in water vapour and carbon dioxide. It is believed that in the accreting process, hydrates, carbonates and nitrogen compounds were trapped in the rocky core. Heating occurred causing the compounds to lose water as water vapour and to form carbon dioxide and nitrogen. These gases escaped through the crustal layer to form the terrestrial planets' secondary atmosphere. This process is called **outgassing**. Confirmation of this can be obtained by looking at the noble gases. These are inert and do not react with the surface rocks and, apart from helium, do not escape from the atmosphere. Therefore you can compare the amount of noble gases present in the atmospheres of the terrestrial planets to that expected to occur for each of the possible mechanisms for creation of the secondary planetary atmospheres. From measurement of the noble gases, outgassing seems the most plausible explanation for the secondary atmosphere (Wayne, 1992). Early in the Earth's history the volatiles were outgassed during three periods of strong heating. The principal mode of heating, and therefore outgassing, would have been the bombardment of the Earth by the remaining planetesimals. This impact outgassing probably took place in the first 0.5×10^9 years (Henderson-Sellers, 1983). As the number of planetesimals declined and the accretion phase stopped, the temperature of the Earth probably fell dramatically, from 1500 to 300 K (Pollack, 1990). Outgassing also occurred during the global internal differentiation phase. This is when the Earth's metallic core formed in the latter stages of accretion. Finally, outgassing occurred by volcanism, perhaps extremely violent, in these early days. At the present time outgassing can occur principally by volcanism or the weathering of igneous rocks. The former mechanism is believed to be the most important.

The Faint Sun Paradox

The secondary atmosphere is not the atmosphere that the terrestrial planets have now. The secondary atmosphere has continued to evolve in response to various changes. These include changes in the planet's rotation rate, changes in the amount of gaseous absorbers and planetary albedo. The Sun is the ultimate thermostat, controlling the temperature of all the terrestrial planets. Thus the most important change for planetary atmospheres has been in the amount of solar radiation received by each planet due to changes in solar luminosity. Furthermore, the total flux or emission spectrum of the Sun has changed over its history. The ultraviolet flux from the Sun was far greater early on than it is now.

The evolutionary history of the Sun is probably better known than that of the Earth. Astronomers have a classification system for stars and the Sun is a typical G2 star born about 5×10^9 years ago. The classification of stars is shown in the Herzsprung-Russell diagram which describes the evolution of a star and its luminosity over its lifetime (Croswell, 1994). The Sun creates energy by turning hydrogen into helium by nuclear reactions deep within its interior. This has led to a steady increase in solar luminosity since the beginning of the solar system. Calculating the luminosity of the Sun reveals that this increase is about 43 per cent (Figure 3.1). This has had a dramatic effect on the climate of the planets.

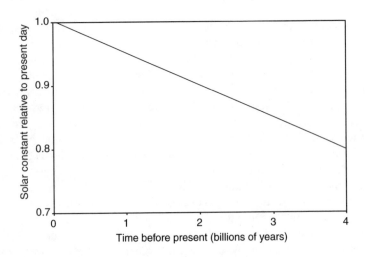

Figure 3.1 Change in solar luminosity over geological time
Source: Kuhn *et al.* (1989)

EARTH

The lower luminosity of the Sun should have meant that two thousand million years ago the Earth's oceans were frozen. All the geological evidence points to the exact opposite. Similarly, the Mars terrain indicates a less harsh climate at this time (Pollack, 1990). It can be established from geological data that liquid water has been present on the Earth's surface for at least 80 per cent of its history (Henderson-Sellers, 1983). The equitable climate of Earth at a time when the solar luminosity was so low is known as the **Faint Sun Paradox**. It seems likely that the main factor to account for this paradox is a stronger greenhouse effect. Carbon dioxide is likely to be the primary greenhouse gas responsible (Kasting and Toon, 1989). This is because of the strong evidence supporting the volcanic outgassing of a secondary atmosphere, and experiments showing that life can evolve under these conditions. To offset the Sun's lower luminosity early in Earth's history it has been suggested that 300 to 3000 times more atmospheric carbon dioxide than at present would have been needed (Pollack, 1990). This is possible through increased volcanic activity and less weathering (the removal of carbon dioxide from the atmosphere by rocks), due to a smaller continental land mass. Henderson-Sellers (1983) states that it is these initial conditions, which allowed the condensation of water vapour, that determined the evolutionary fate of Earth. Probably a large, but not global, ocean occurred. This allowed a hydrological cycle, similar to today's, to operate. The large oceanic extent and the presence of a considerable atmosphere may have protected the Earth from any violent climatic change as the poleward transport of heat would have been well established.

VENUS

As Table 3.1 shows, Venus is much more of a sister planet to Earth than Mars. It is almost the same size and mass of Earth with a similar composition and gravity. Venus probably began with the same volatiles as Earth. Yet now its atmosphere is predominantly carbon dioxide, with a cloud top temperature of $-33°C$ and a surface temperature of $480°C$ and a surface pressure of $90\,000$ mb. Of course Venus is closer to the Sun than Earth so we might expect it to have a higher temperature. But how has it followed such a different evolutionary path? Its closer position to the Sun would have led to a rapid outgassing of carbon dioxide and water vapour and formation of an atmosphere. The low solar luminosity then would have allowed the formation of oceans, although they would have been hot, probably not much below the boiling point of water. The clouds in the atmosphere would reflect solar radiation away cooling the planet but would also trap infra-red radiation warming the surface. Overall, the cooling effect would have dominated. This would have increased the planetary albedo but the temperature probably remained high. Water would easily evaporate from the hot oceans into the atmosphere.

This combination of hot oceans and moist atmosphere is known as a moist greenhouse (Pollack, 1990). The high temperatures would allow the water vapour to reach high into Venus's atmosphere. Here it could be split into its constituent parts of molecular oxygen and hydrogen by ultraviolet sunlight (**photodissociation**). This process would have been aided by the increased amount of solar ultraviolet flux at this point in the Sun's evolution. Photodissociation of water would have resulted in the loss of hydrogen to space but oxygen would be left in the atmosphere. The oxygen may have been taken up by surface weathering along with carbon dioxide. The oceans would keep supplying water vapour to the atmosphere which would be continually lost. Eventually so little water was left it could no longer liquefy and the oceans disappeared. At this point any carbon dioxide outgassed would remain in the atmosphere as no weathering could occur to return it to carbonate form. The temperature kept increasing with a runaway greenhouse effect until eventually it stabilized at a high temperature. Unlike Earth, on Venus, most of the carbon dioxide now resides in the atmosphere. The presence of liquid water on Earth allows a similar amount of carbon dioxide to reside in the oceans and rocks.

MARS

Mars is much smaller than Venus or Earth therefore although tectonic activity and outgassing occurred it is likely to have been at a much smaller level. This in itself does not explain the tenuous atmosphere Mars has today. Mars has lost a considerable amount of atmosphere. It has been suggested that most was lost to space, with photochemistry playing an important role (Wayne (1992) quoting McElroy *et al.* (1977)). It seems likely, however, that most has been incorporated into the surface with little loss occurring by photodissociation (Henderson-Sellers, 1983). Mars, being farther away from the Sun and with the low solar flux, would have experienced very low temperatures. This would have initially hampered outgassing with mostly carbon dioxide being released. As the carbon dioxide atmosphere formed, a greenhouse effect would have been established. The temperature of Mars would then have risen, allowing increased volcanic activity and further outgassing of carbon dioxide. Liquid water would have formed allowing weathering to occur and for carbonates to form.

It is not its distance from the Sun that sealed Mars' fate, but its size (Kasting and Toon, 1989). As it is a small planet, Mars cooled much more quickly than Earth. Consequently, volcanic activity declined and carbon dioxide could not be returned to the atmosphere. Low temperatures also led to freezing out of carbon dioxide at the poles increasing the albedo. This would have compensated for the increasing solar flux. The temperature of Mars gradually became too low to allow liquid water to remain. It has been speculated that the vast majority of the carbon dioxide remaining in the Martian atmosphere was then absorbed onto the tiny grains of sediment that make up the Martian surface layers (Pollack, 1990). The remaining carbon dioxide atmosphere of present-day Mars provides a greenhouse effect but the atmosphere is so tenuous that the temperature remains stable but very low.

Goldilocks

The importance of carbon dioxide as a naturally occurring greenhouse gas cannot be stressed enough. Together with other factors the amount of atmospheric carbon dioxide probably determined the climatic fate of Venus, Mars and Earth. It was carbon dioxide which in the early days of Earth's existence kept it from freezing over while the solar luminosity was low. Looking at the current atmospheric composition of Earth, however, it is quite striking how it differs from the carbon dioxide-dominated ones of Venus and Mars. It is nearly universally agreed that this difference is due to life on Earth. The story starts with the Sun–Earth distance. A mere 5 per cent closer and the Earth would have lost all its oceans by now (Schneider and Londer, 1984). Next is the mass; the Earth was large enough to allow enough tectonic activity and outgassing. These characteristics allowed Earth to develop an atmosphere in which the amounts of carbon dioxide and water vapour were just right at the right time. This atmosphere could then develop a moist circulation allowing the redistribution of heat from the equator to the poles. Planet Earth is often referred to as the Goldilocks planet, from the story of Goldilocks and the three bears (Pearce, 1989). When Goldilocks enters the house of the three bears she tastes the porridge, one was too hot (Venus), one was too cold (Mars) but one was just right (Earth). With the right starting characteristics not only did the Earth have the right temperature but it was also able to maintain it long enough for complex life to evolve. Once life appeared, it began to modify the atmosphere, by just how much will be considered in the next section.

It has always been speculated that at some point in the past all three terrestrial planets, Venus, Earth and Mars have evolved life. Some geological evidence suggests that life began far earlier in the evolutionary history of Earth than was once suspected. It may be that life is an inevitable consequence of planetary formation. However, the evolution of complex lifeforms appears to require liquid water (Appel and Moorbath, 1999). Venus may well have gone through an Earth-like phase, when life could have existed, before becoming hot and dry (Pollack, 1971 and 1979). However, the hot moist ocean phase, if it occurred, only lasted a few hundred million years and may still have been too hot for life to evolve. The possibility of life on Mars has fascinated humankind. Recently it has been claimed that fossilized forms found in a meteorite from Antarctica provide evidence of life on Mars. This claim, however, is heavily contested. It does seem quite likely that Mars had liquid water at some early stage of its evolutionary history and perhaps a series of oceans (Morton, 1998). This period may even have been long enough to produce life. If life exists now on Mars it is most likely to be subterranean. With the loss of most of its atmosphere Mars is now a cold tundra planet. How solar systems form is still a matter of debate so whether the characteristics of Earth are a fortuitous occurrence or commonplace is unknown but life did form. The question now is how over the lifetime of the Earth, with increasing solar luminosity, has the temperature been maintained such that life could continue to evolve. Essentially this becomes a question of how the amount of carbon dioxide in the atmosphere has been controlled.

Given that carbon dioxide is such an important greenhouse gas it is essential that we understand how it has been removed at a rate which is just sufficient to maintain the Earth's temperature at between 15°C to 30°C. There are two main theories for how this occurred (Pearce, 1989). The generally accepted theory is the geochemical model. The second theory is a biological model which has received world-wide attention and captured the public's imagination.

The Geochemical model

This model is based on the carbonate-silicate cycle which forms part of the carbon cycle as discussed in Chapter 2 (Walker *et al.*, 1981). Rain removes carbon dioxide from the atmosphere as dilute carbonic acid. At the surface this acid combines with the silicate materials which converts the carbon dioxide into bicarbonate. These bicarbonates are then dissolved in water and carried to the oceans. In the oceans the carbonates are incorporated into the shells of sea creatures. Eventually these will be laid down as sedimentary rocks. To return carbon dioxide to the atmosphere requires tectonic activity which Earth still has in abundance. The sedimentary rocks on the sea floor are carried out by the spreading ocean floor to the edges of the continental plates. Here the rock slides down into the Earth's interior to be subjected to extremes of temperature and pressure. This releases the carbon dioxide to be outgassed at mid-ocean ridges and volcanoes. It is important to understand how this carbon-silicate cycle can regulate the Earth's climate. First, the transformation into bicarbonate depends on the temperature and second, the processes rely on precipitation and run-off which are also temperature dependent (Kasting and Toon, 1989). In warm wetter climates more carbon dioxide is removed from the atmosphere due to the increased precipitation, therefore reducing the greenhouse effect. In cold climates, when water is frozen, there will be less precipitation to return the outgassed carbon dioxide to the surface. Any volcanic activity will allow carbon dioxide to build up in the atmosphere increasing the greenhouse effect. It is this slow process of chemical weathering and return by tectonic activity that is the accepted explanation for the long-term decline in atmospheric carbon dioxide compensating for the growing luminosity of the Sun. During the early Earth and the Faint Sun, tectonic activity would have been high and weathering less due to smaller continental land masses. This would have lead to greater amounts of atmospheric carbon dioxide, and a larger greenhouse effect and so to temperatures high enough to maintain a liquid water environment. The smaller Mars quickly converted its carbon dioxide atmosphere to rocks and its quickly declining volcanism, in contrast to the larger Earth's slower cooling rate, meant the cycle of returning carbon dioxide to the atmosphere was lost. The Earth has cooled slowly and still has a geological cycle of carbon dioxide. The evolution of life, however, has modified these geochemical cycles. One of the most important ways is in the growth of the oxygen content of the Earth's atmosphere.

In the early atmosphere, before the formation of life and in its early stages, the amount of oxygen in the atmosphere was controlled by chemical processes. Water vapour released by volcanic activity into the atmosphere would be photodissociated by sunlight into hydrogen and oxygen. This would have occurred much more readily in the early atmosphere as there would be no ozone layer to shield the Earth from the high energy ultraviolet radiation from the Sun. The hydrogen would escape to space, while some of the oxygen remained in the atmosphere. A proportion of oxygen would be taken up through the oxidation of surface rocks in the weathering process. As life evolved, photosynthesis became an increasingly important contributor to the atmospheric oxygen content. Gradually oxygen levels rose, first in the atmosphere and surface ocean water and then in the deep ocean. In the atmosphere this allowed the formation of an ozone layer. The ozone layer protected both life and water vapour in the lower atmosphere from ultraviolet radiation. While volcanic activity, photodissociation and weathering all still play a part in the control of the oxygen content of the atmosphere, their role is now extremely minor. The control of atmospheric oxygen content is now predominantly a biological one, similar to the carbon cycle, as introduced in Chapter 2.

In summary, the geochemical model places emphasis on the geochemical cycles. It is these that regulate the amount of carbon dioxide in the atmosphere and therefore through the greenhouse effect the temperature. The biosphere has modified these cycles to produce a predominantly oxygen–nitrogen atmosphere.

The biological model

The biological model was postulated by Jim Lovelock in the 1970s (Pearce, 1989). It is based on the idea that the Earth is 'alive' and is regulated by the biosphere for the biosphere. Lovelock named the idea **Gaia** after the Greek Earth Goddess (Lovelock, 1988). The original idea was based on phytoplankton which are found all over the globe in oceans. They consume carbon dioxide which the ocean has absorbed from the atmosphere. When they die their carbonate skeletons fall to the bottom of the ocean to form carbonate rocks. Lovelock proposed that in a warmer world there would be more of these plankton, they would require more carbon dioxide, which would reduce the carbon dioxide available in the atmosphere and so reduce the greenhouse effect which would cool the Earth so reducing plankton numbers. This original idea has flaws, phytoplankton were not around in the early Earth and yet it still maintained a reasonably constant temperature. Also, there is little evidence that the number of phytoplankton depends on temperature. It is nutrients which limit phytoplankton numbers. Lovelock suggests that in a warmer moist world with greater rainfall, nutrients would be more easily washed into the sea thereby encouraging greater numbers of phytoplankton (Pearce, 1989). Lovelock has developed his ideas arguing that in the current world the micro organisms in the soil control weathering and in the ocean they control the intake of carbon dioxide. The important point is that in this model not only has the biosphere modified the atmosphere to produce an oxygen–nitrogen based one but it also regulates the amount of carbon dioxide.

A major early criticism of Gaia was that in order for the biosphere to regulate the Earth's temperature in this way it would be required to have foresight. To answer his critics Lovelock and a colleague developed a model called **Daisyworld** (Watson and Lovelock, 1993). Daisyworld is a planet with no clouds and its atmosphere has a negligible greenhouse effect, i.e. the atmosphere is almost transparent in the infra-red. The only things on the planet are two species of daisy, one dark and one light, the former reflects less light than the bare ground and the latter more. They tend to be referred to as black and white daisies. Only four mathematical equations are used to govern Daisyworld. However, these form a system of non-linear multiple feedback loops, which are difficult to solve. Watson and Lovelock (1983) showed that when the daisies are growing at a steady rate then the local temperatures of black and white daises are stable. This means that given a perturbation, the daisies will respond by restoring the local temperatures to pre-fixed values. They also showed that the steady state planetary temperature would actually decrease in response to increased solar luminosity. Watson and Lovelock (1983) then used a computer to integrate the equations to steady state for a few specific cases. Their findings showed that the introduction of daisies allowed a steady temperature to be maintained over a wider range of solar luminosity than if there were no daisies (*see* Fig. 3.2). There is a strong environmental feedback at work in this model. Black daises are warmer than white and therefore prefer a cooler climate, yet an increase in black daisies warms the planet, limiting their growth. From this it might be expected that the model will be stable, resisting any perturbation.

Watson and Lovelock (1398) address this problem by causing a major change in the Daisyworld climate. Clouds appear over black daisies. As clouds are white they cool the planet, so more black daisies means a cooler planet. On running the model white daisies become extinct because they are less able to survive, black daisies flourish and maintain the balance. The main

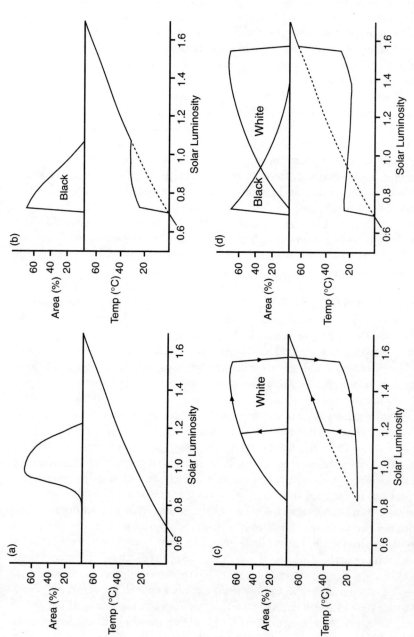

Figure 3.2 The results for Daisyworld showing the response of different coloured daisies and temperature to increasing solar luminosity: (a) for grey daisies, the same colour as the planet (b) for black daisies only (c) for white daisies, note that decreasing the solar luminosity does not lead to the same temperature profile and (d) for both black and white daisies
Source: Watson and Lovelock (1983)

conclusion Watson and Lovelock (1983) give is that regardless of the direction of feedback, the model always shows greater stability with the daisies than without them. Although they state extrapolation to the real world is tenuous, they point out that the biota may have a substantial influence on the Earth's temperature via the abundance of greenhouse gases in the atmosphere. The important concept is that biota influence the temperature. This can be extended to the idea that biota may play a part in the stabilization of the Earth's temperature. Charlson *et al.* (1987) have suggested that marine phytoplankton play this role. These phytoplankton produce a gas called dimethyl sulphide (DMS). The gas reacts with oxygen in the atmosphere to form sulphate aerosols. These aerosols provide the perfect surface for condensation to take place and therefore allow clouds to form. Larger numbers of cloud condensation nuclei lead to a change in the clouds' properties. They have a higher albedo and thus reflect more incoming solar radiation and have a cooling effect. In this way the phytoplankton can regulate the Earth's temperature and climate. Warmer temperatures seem to lead to greater DMS production (Charlson *et al.* 1987). This would result in brighter clouds and cooler conditions. The DMS production would then decrease and so less bright clouds would occur allowing warming. Thus the global climate could be regulated. It has been suggested that the phytoplankton developed such a system to allow them to use clouds and the associated air currents to disperse about the globe conferring an evolutionary advantage (Hunt, 1998). From these ideas a leap can be made to the concept of Gaia, of a biosphere and its physical environment so closely coupled that they can be considered as one organism.

Gaia remains a controversial theory and, although acknowledged as a great contribution to the scientific debate, arguments continue. Some of these arguments focus on the scientific credibility of Gaia. For instance Budyko (1986) is highly critical of Gaia and the idea that the biosphere can have a global influence, basing his arguments on the precariousness of life on Earth. Life can only exist in particular conditions which are far from common even on the Earth. Thus the flexibility of the biosphere seems limited, and its longevity even more surprising. Budyko argues that the use of the idea of a superorganism is only justified if there is a great deal of stability within the system both to internal and external events, but particularly to the latter. He points out that organisms affect their environment so as to improve the conditions for their existence, but only in local ecological niches over comparatively short time-scales having little global impact. Budyko concludes that 'It seems to us that the authors of the "Gaia hypothesis" have not based their hypothesis on firm grounds'. Life on Earth, however, has been found in extreme conditions. In the sunless deep oceans living in hyperthermal vents where the temperatures reach 350°C and in rocks in the dry arid wastes of Antarctica where temperatures rarely go above freezing. The key to life is always liquid water but the range of temperatures over which life can survive are perhaps far greater than that envisaged by the stability of Gaia.

Many of the other arguments are far from scientific and have more to do with how Gaia has been interpreted from both a political and theological standpoint. Gaia has been strongly associated with the green movement. However, both sides of the political 'fence' have used Gaia to bolster their arguments. Both rest on the idea of a biosphere being central to the operation of an equitable global climate. The green case concludes that if we destroy the biosphere either through direct actions such as tropical deforestation or indirect ones such as global warming (we will see exactly how later), then we will have destroyed the regulator of our climate and further deleterious climate change will be inevitable. The counter-argument is that as we have a biosphere then it doesn't matter what humans do as the biosphere will regulate it and stop any climate change for the worse. When we talk about the biosphere we are also considering the notion of biodiversity, that is the multitude of various different organisms that currently live on Earth. As humankind requires more and more space, so the rate of **extinction** of species has increased leading to decreasing biodiversity. Tropical rainforests are a whole ecosystem and green fears are that if we continually destroy these and other ecosystems then we seriously impair the viability of the biosphere. Those countering this argument point out that biodiversity is not necessarily a prerequisite for a biosphere, after all Daisyworld relied on just two types of daisies! However, this is a dangerous route to follow because Gaia might just decide that it would be better off without its problem child human beings and let some other species live on and evolve to control the biosphere. There is nothing in Gaia to say that it is us that must survive (Schneider and Londer, 1984).

The importance of Gaia from a theological viewpoint is that it presents us with a whole Earth view. Humans and science are not separate from the Earth but part of it. The theological debate has been examined by Houghton (1994). Principally this is through the Christian sentiment of 'stewardship'; that is that the Earth has been given to human beings to look after and care for, on behalf of God. Houghton is quick to point out that many other religions such as Islam, Hindi and other ancient religions have similar notions. The ideas of stewardship have powerful supporters (Gore 1992; Houghton, 1994) and stems from the Judaeo-Christian tradition of human dominion over the Earth (Gore, 1992; Portney, 1992). However, this is not the only interpretation of dominion. For some it means that the environment should be protected from economic market forces at all costs whereas others hold exactly the opposite view. For them dominion means that all of the Earth's resources are there to be used up in order to better humankind's position (Portney, 1992). Gaia has therefore become more than a scientific theory. Perhaps this is what has held back a more widespread acknowledgement of the potential influence of the biosphere on climate. It certainly provides some insight into how science is socially constructed and can be used to support a variety of beliefs.

The geological climate record

It was the geological evidence of liquid water on Earth that presented the Faint Sun Paradox. How can one reconstruct the past geological climate (for the geological time-scale, *see* Figure 3.3)? There are no direct measurements of the temperature and the climate is deduced from data that reflect the natural conditions of the past. As these data are not direct measurements of the climate they are called **proxy data**. Oxygen-18 isotope analysis, sediment formation, rock weathering, formation of water reservoirs and the existence of living organisms are all dependent on atmospheric factors and thus the climate. There are problems associated with this approach. The assumption has to be made that the relationship between natural phenomenon and the climate was the same then as it is now. By using several paeleogeographic indicators, however, this problem can be minimized due to the variety of natural process dependent on climate. The climate record also requires dating. The main sources of data and dating methods are reviewed below and provide a flavour of the problems paeleoclimatologists face when trying to derive the climate of the Earth's distant past (for greater detail see Condie and Sloan (1998), Dawson (1992) or Huggett (1991)).

Dating the past

COMPARATIVE DATING

In comparative dating, also called relative dating, rocks, sedimentary layers and fossils are examined to see if they are similar to those found in other areas. If they are, then they were probably formed at the same time. However, this does not give an actual date. Also for Precambrian rocks, which lack fossils, even comparative dating is difficult. In the early days of geological science, however, this was really the only way to date findings. It led to the naming of the geological periods and time sequence (*see* Fig. 3.3).

ABSOLUTE DATING

Absolute dating relies on radioactive dating. Every atom is composed of a nucleus of protons and neutrons surrounded by a cloud of electrons. A proton has a positive charge, an electron a negative charge and a neutron no charge. The number of protons is unique for each element and is called the atomic number. The number of protons and neutrons is the atomic weight or mass. Different isotopes of an element have the same number of protons but a different number of neutrons. Some isotopes are stable but others disintegrate to form atoms of a different element releasing energy in the process. This rate of disintegration is fixed and does not vary, thus once a radioactive element is created it starts to disintegrate. This rate of decay is often stated

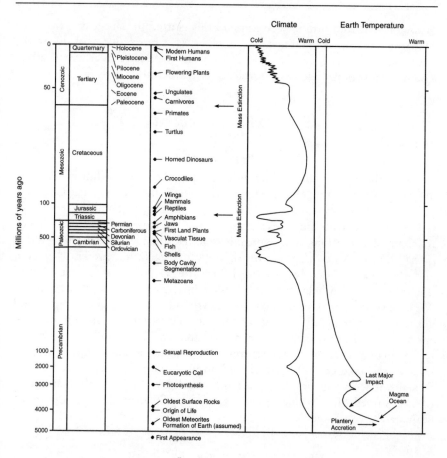

Figure 3.3 Geological time-scale with an overview of the climate and temperature of the Earth
Source: After Condie and Sloan (1998)

as **half-life** - the time required for one-half of the original number of radioactive atoms to decay. By looking at the ratio of radioactive isotopes or parent elements, to the new element (daughter) formed and by knowing the half-life we can date the rocks. Carbon-14 is formed when neutrons from cosmic rays interact with nitrogen in the Earth's atmosphere. The carbon-14 then decays back to nitrogen with a half-life of 5730 years. Therefore **radiocarbon dating** is only useful for, at most, the past 70 000 years. In contrast, Rubidium-87 decays to Strontium-87 with a half-life of 48.8 billion years which makes it most useful for dating Precambrian rocks and can be used for ages over 100 million years old. This dating method is the one of the most important and widely used in geology. Another important isotope is Potassium-40. This isotope decays to form Calcium-40 and Argon-40; 89 per cent of the Potassium forms calcium, the remainder forms Argon. Argon-40 can be easily distinguished from ordinary Argon and this is used to date fossils from the Mesozoic era, that is the dinosaur era. However, a problem occurs due to the inertness of Argon. It can be removed from the rock, particularly by heating, leading to a younger age being ascribed to the rock than it really is.

The climate record

SEDIMENTARY DATA

Knowing the type of climatic conditions required to form sedimentary rocks allows geologists to infer past climates from these deposits. Chemical weathering of rocks is more rapid in hot humid climates, and the volume of sedimentary deposits increases with wetness. Coal formation is closely linked to climatic conditions and typically requires abundant vegetation and a moist climate. Much depends on the type of vegetation present, however, and the link is not always straightforward. Deposits of limestone and dolomites can also help to establish the climatic conditions of water reservoirs. **Varved** clays (sedimentary layers) around the margins of continental glaciers can reveal information about seasonal variations. In summer and spring the glacier melts and the runoff is greater. This enables larger debris to be carried away by water streams which forms a coarse light-coloured layer at the lake edge. In winter and autumn the runoff is small and the deposits are consequently finer, forming a dark layer. The stratified structure of the varved clays allows the length of their formation period to be evaluated.

GEOMORPHOLOGICAL EVIDENCE:

Geomorphology is widely used to study atmospheric precipitation and ice cover formation. The position of the coastline, i.e. sea level is used to measure the extent of continental glaciation. Surface relief variations are used to indicate glacier development and therefore the extent of ice cover. The thickness of the ice and the direction in which the ice moved can also be determined in some cases. Fluctuations in the level of land-locked seas allow water flow to be evaluated. Erosion is strongly influenced by climate. River terraces where the river cuts into the underlying rock can also provide information on geological climates. The river flow and hence the terraces are dependent on climatic conditions and can provide data on sea level, glaciation and climatic change.

DATA ON FOSSIL FLORA AND FAUNA

The geographical distributions of plants and animals, but particularly vegetation, are reliant on climatic conditions. Smooth-edged leaves are indicative of tropical temperatures whereas leaves with a jagged edge predominate at lower temperatures. **Dendrochronology** or the study of tree rings can also be used. Fossilized tree rings indicate a climate with seasonal variations whereas fossilized trees without growth rings indicate a tropical climate. Variations in the annual growth layers reflect short-term climatic variations, and the structure is indicative of general climatic conditions. However, dendrochronology is more useful in the recent climatic past and will be discussed in detail in the next chapter.

PAELEOTEMPERATURES

The O^{18}/O^{16} ratio in mollusc shell and other marine organisms reflect the temperature in which they live. This ratio can also be ascertained from accumulated snow and ice. There are three isotopes of oxygen, 16, 17, and 18. Each isotope has the same number of protons (8) but the number of neutrons vary. O^{16} has 8, O^{18} has ten. O^{17} occurs in too small quantities to be of importance. In water 0.2 per cent of the H_2O molecules are made from O^{18}, these are heavier than the 99.8 per cent of H_2O made from O^{16}. The heavier molecules find it harder to evaporate from liquid water. So water vapour in the atmosphere is deficient in O^{18} compared to sea water. The condensation process, to form rain and snow, also separates the oxygen isotopes. The amount of separation is determined by temperature because both evaporation and condensation are temperature dependent. Water vapour composed of O^{18} condenses more easily and so, as condensation occurs, the water vapour in the atmosphere contains less O^{18}. If the temperature continues to fall, and condensation persists, the rain and snow which go to form the ice layer contain less and less O^{18}. Ice formed by the accumulation of snow therefore contains a record of the atmospheric temperature. By looking at the oxygen isotope ratios in ice cores from the polar regions, paeleotemperatures can be obtained. Generally, the lower the O^{18} content of the ice, the lower the temperature.

As an **ice age** develops, more ice is deposited on the land, there is less water in the ocean and the concentration of water formed from O^{18} molecules rises. The ratio of O^{18} to O^{16} in the oceans rises as an ice age persists. The oxygen isotope ratio is also dependent on the depth and temperature of the sea water. Sea animals with shells which are formed from calcium carbonate, called foraminifera, reflect the concentration of O^{16} to O^{18}. Thus the shells reflect the oxygen ratio in the sea at the time the animals were alive. The animals die and the shells fall to the seabed to form sediments. The sediments can then be removed as cores from the ocean beds and the oxygen isotope ratios of various layers can be studied. There are difficulties in deriving the temperatures from these isotopic concentrations. Evaporation over the globe is not uniform thus neither are the concentrations of O^{16} to O^{18}. However, ocean currents stir up everything and in a few thousand years the sediment is sufficiently mixed so as to provide uniform concentrations but this does mean that the time-scale is not very detailed. The main problem arises from specifying the variations in the oxygen isotopes due to changes in water temperature and those due to changes in the isotopic composition of the sea water. To do this, the oxygen isotope ratio during an ice age has to be calculated. This requires the ice volume and sea water volume to be known. It is estimating these which has caused the greatest difficulty in interpreting oxygen isotope cores. As the temperature of the deep ocean water changes little, even between glacial and interglacials, foraminifera from here reflect mostly the isotopic composition of the sea water. In contrast, foraminifera living in the surface waters reflect oxygen isotope changes due to surface water temperatures (Dawson, 1992).

Geological climate change

There are many slow long-term changes, apart from the change in solar luminosity, that affect the climate of the Earth. Prior to 3×10^9 years ago the Earth was mostly covered in water (*see* Fig. 3.4). Around 250 million years ago there existed one large supercontinent called Pangaea. This broke up about 180 million years ago into two continents called Gonwanaland and Laurasia. Gowanaland is believed to have been composed of South America, Africa, Australia, India and Antarctica. Laurasia was made up of North America and Eurasia. At the beginning of the Cenozoic, 65 million years ago these two large continents stared to break up to form the continents as they are now. The formation of continents and mountain building will have affected the albedo, thereby changing the radiation balance and atmospheric circulation. The rotation rate of the Earth has slowed over time due to tidal friction with

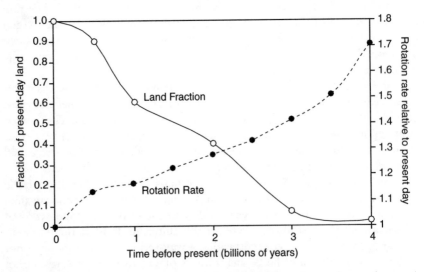

Figure 3.4 The changing rotation rate and continental mass growth over the lifetime of the Earth
Source: Kuhn *et al.* (1989)

the Moon. Heat transport from the equator to the poles, in both the atmosphere and ocean, is dependent on the rotation rate so any changes will affect the climate (*see* Fig. 3.4). Kuhn *et al.* (1989) suggest that the Earth may well have rotated up to 70 per cent faster 4×10^9 years ago than it does today. That would mean a day length of 14 hours. During the Silurian period (420 million years ago) a year was 421 days and the length of the day was 21 hours. This would have reduced the temperature difference between night and day and may have influenced biodiversity. The experiments of Kuhn *et al.* (1989) with a simple climate model, imply that the change in rotation rate is an important factor affecting the climate over geological time periods.

In summary, changes in the long-term geological climate are attributed to plate tectonics and astronomical changes. These influence the Earth's surface temperature through changes in the radiation balance, by changing the albedo or the greenhouse effect. Plate tectonics determine the size and position of the continents and also control mountain building, volcanic activity and sea level changes. The land/sea ratio affects the albedo and tectonic activity atmospheric composition. Plate tectonics also governs the size and shape of the oceanic basins and therefore the ocean currents. This in turn affects the general circulation and distribution of heat from the equator to the poles. Astronomical changes include the long-term increase in solar luminosity and changes in the Earth's orbital characteristics. Variations in the Earth's orbit are thought to be responsible for the quaternary glacial cycles and will be discussed later in the chapter. There are, however, other times in the more distant geological past when the Earth has had considerable ice sheets.

The popular perception of climate change is that of the ice ages, however, the term ice age is used rather loosely. Ice ages in geological terms occur over time-scales of 100 million years. Quaternary ice ages occur on 10 000–100 000 year time-scales, and are termed glacial cycles. The Earth is currently in an **inter–glacial period**. To confuse matters even further there is the Little Ice Age – a period of pronounced cooling in Europe which began in late medieval times. This cooling has a period in the order of 100 years. The mechanisms responsible for these various 'ice ages' are different, not just because of their intensity, but because of the different time-scales involved. This is a very important point. Any mechanism must have the same period as the cycle if it is to explain the change in climate.

The geological climate record suggests periods called 'greenhouse' and 'icehouse' Earths. These are times when the Earth was supposedly completely ice-free at the poles (greenhouse) and then with considerable ice in the polar regions (icehouse). Therefore in geological terms the Earth is currently in an ice age. Generally, warmer periods in the geological past are associated with increased levels of carbon dioxide and colder periods with decreased levels of carbon dioxide. However, it is unclear whether the change in carbon dioxide precipitates the climate change, exacerbates an existing change or is a response to a change. High carbon dioxide levels seem to be necessary to explain most of the ice-free times on Earth. Studies using computer models have supported the notion that if there was a pole-centred supercontinent, then snow-free conditions might have occurred, at summer time, even without high levels of carbon dioxide. However, more recent sophisticated modelling studies (Crowley *et al.*, 1993) have shown that the real-world geography may not have allowed a snow-free polar supercontinent to exist. Other computer model studies have, on the whole, shown that such polar supercontinents initiate glaciations.

The geological ice ages are shown in Figure 3.3 along with the geological time-scale; usually six ice ages are identified. In the past one million years the ice cap has reached down to lower latitudes four times. During the Mesozoic era no major ice advances are recorded. In the Early Permian, Late Carboniferous and Late Ordovician (Paeleozoic era) there were numerous glaciations. There was another period of glaciation in the Early and Late Proterozoic (Pre-Cambrian). Ice ages appear to have occurred at regular intervals in the past, about every 155 million years with a range of 140 to 170 million years (Huggett, 1991). There is probably no one cause to account for these ice ages but the position of a polar continent appears to be an important feature. Glaciations in the Pre-Cambrian are associated with a reduction in carbon dioxide in the atmosphere. There are several theories for how the carbon dioxide was removed from the atmosphere (Condie and Sloan, 1998). One suggestion is that the carbon dioxide was taken up by photosynthetic micro-organisms. However, this does not provide an explanation of how the glaciation ended. A more likely theory is that the glaciations at this time were due to the formation and break-up of a supercontinent, the changes in carbon dioxide levels relying on the geochemical cycle. A growth of land mass would lead to an increase in weathering and a reduction of carbon dioxide in the atmosphere. This would result in a reduced greenhouse effect and lower temperatures leading to glaciation. The glaciation ends because ice would cover the rocks decreasing carbonate sedimentation due to weathering. Therefore carbon dioxide uptake would decrease until finally the amount of carbon dioxide in the atmosphere would rise. This would increase the greenhouse effect and lead to higher temperatures. It would appear that a continent over either the south or north pole can lead to world-wide ice ages. The Paeleozoic glaciations appear to be in response to a pole-centred continent. While the late Devonian and late Ordovician rely on a south pole continent, the late Permian glaciation can be explained by a north pole-centred continent (Huggett, 1991). Similarly, the most recent glaciations appear to be associated with land masses near the polar regions and this may be a prerequisite for glaciations. It could also explain the switch from an icehouse to greenhouse mode as well as the periodicity of the variation. However, investigations of the most recent glaciations suggest that changes in the seasonal distribution of solar radiation may have an important impact on ice build-up (Huggett, 1991).

While looking at geological climates provides an interesting insight into climate change and the workings of the greenhouse effect, the time-scales involved are just too long to explain global warming. The geological changes are on the scale of planetary evolution and they will not be discussed further. Condie and Sloan (1998) or Rogers (1993) give a fuller account. However, studying climates in the geological past with radically different carbon dioxide contents to the present day can help in understanding the effects of increasing carbon dioxide on the current climate. These ideas will be considered in Chapter 5, but now we will turn our attention to more recent events and the quaternary glacial cycles.

Mechanisms for quaternary glacial cycles (ice ages)

The most recent history of the Earth, termed the quaternary (*see* Fig. 3.3) has seen frequent and cyclical advances of the ice caps. There are no strict definitions of what constitutes a quaternary ice age such as a critical temperature or latitudinal limit to the ice sheet. An interglacial period has been roughly defined as when oak trees and deciduous trees are commonplace in Europe. A glacial epoch begins when these types of trees disappear from Europe (Imbrie and Imbrie, 1979). Over the past 700 000 to 800 000 years there have been six to seven glacial advances and interglacial retreats. Glacial advance and retreat appears to occur simultaneously in the Arctic and Antarctic. Theories about the cause of the repeated glaciation during the quaternary period abound. They fall into two broad categories, those which rely on orbital changes to promote an ice age and those which do not. The latter are mostly concerned with an internal perturbation of the climate, or part of the natural climate cycle.

Internal mechanisms

Several mechanisms revolve around the positive ice-albedo feedback which propose that an increase in snow and ice will lead to a cooling of the Earth and a further increase of snow and ice. An example of this is the theory suggested by Wilson, in 1964. He proposed that periodic surges of Antarctic ice might initiate quaternary ice ages (Imbrie and Imbrie, 1979). The Antarctic ice sheet accumulates snow on the surface, becoming thicker and heavier. Normally this mass of snow flows slowly to the edges of the ice sheets. Then the ice breaks off in the form of icebergs. If the snow continued to accumulate at the bottom of the ice sheet, the pressure would increase and this would eventually cause the ice to melt. This would allow the movement of the entire Antarctic ice sheet to increase and the ice sheet would surge out into the oceans. As it spread and broke up it would reflect more radiation into space, cooling the southern hemisphere. Wind and ocean currents would then spread the cooling northwards and finally ice would form in the northern hemisphere. Glaciation would cease because the outward surge would eventually cool the lowest layer of the Antarctic ice sheet causing it to freeze and the surge to stop. Evidence for an Antarctic ice sheet surge include sharp drops in the levels of O^{18} in deep-sea cores. Although such variations do exist, they do not provide conclusive evidence as they could be the result of other processes. Further evidence for such an ice surge would come from a sudden rise in sea level just before the onset of glaciation, instead a slow decrease in sea level in the build-up to a glaciation has been found (Imbrie and Imbrie, 1979).

Volcanism and geomagnetism have also been put forward as theories of ice ages. The large volcanic eruptions responsible for flood basalt, such as the Deccan Traps in India, may well have a long-term (geological) effect on the climate, perhaps even be responsible for mass extinctions. Volcanic eruptions, as will be shown in the next chapter, may also affect climate on short time-scales. There is little evidence that volcanic eruptions can cause quaternary glacial cycles. It has been suggested, however, that they may accelerate an already declining climate (Dawson, 1992). A record of the Earth's magnetic field is preserved in iron-rich lava as it forms. The Earth's magnetic field can switch polarity from north to south and south to north. The last time the Earth's magnetic polarity switched was about 700 000 years ago. Also the magnetic field is not constant and the position of the pole jumps about. These jumps occur at intervals of about 30 000 years, which is significant for the consideration of ice ages. But how could such a change affect climate? First, changes in geomagnetic intensity have been suggested to affect surface temperatures because the flux of cosmic rays from the Sun into the upper atmosphere increases as the magnetic field decreases. Also the Earth's magnetic field may affect the climate because the field may have some steering effect on the circumpolar circulation. Although neither method can be proved to work, neither method can be disproved.

Oceanic circulation patterns and surface temperatures have changed with Pleistocene glaciations. There are meridional migrations of the subtropical and subpolar water mass boundaries in the north-east Atlantic. Dynamic circulation changes also occur. The North Atlantic current, part of the oceanic conveyor belt, appears not to have flowed during the last glacial maximum over North America (called the Wincosin maximum). Instead a separate anti-clockwise cold gyre occurred north of about 45°N. The idea that the ocean has two circulation modes, one promoting glaciation and the other inter-glacial is well founded. The sawtooth pattern seen in O^{18} records of (Broecker and Denton, 1990) deep-sea ocean cores has been interpreted as the preservation of one mode while glaciation advances to a maximum, followed by a rapid change of ocean structure and accompanying deglaciation. Whether changes in oceanic circulation are a cause or an effect of glaciation is not quite clear. The next section will show, however, that modifications to the oceanic conveyor belt are an important factor in current theories explaining the onset of quaternary glaciations.

Orbital variations

In 1787 a Swiss minister suggested that the Earth had been subjected to extensive glaciations. He and others were dismissed, their ideas rejected on the grounds that the boulders they based their evidence on were not moved by glaciers but Noah's flood. It was Agassiz in 1837 who provided the first detailed hypothesis and established the theory for repeated glaciations of the Earth. In 1842 Adhémar proposed that astronomically driven changes in the intensity of the seasons might trigger such periodic glaciations (Imbrie and Imbrie, 1979). Anything which affects the amount of solar radiation reaching the Earth's surface will potentially have an effect on climate. Various aspects of the orbit will affect the amount of solar radiation reaching the Earth and thus should affect the radiation balance and the climate. There are basically three components affecting the Earth's orbit as can be seen in Figure 3.5.

OBLIQUITY

Obliquity is a measure of the tilt of the Earth's axis. It is the angle between the plane of the Earth's rotation called the ecliptic (this is the path the Sun appears to describe on the celestial sphere) and the celestial equator (the Earth's equator drawn out on the celestial sphere). The obliquity varies from $22°$ to $24.5°$ over 40 000 years. The current obliquity is $23.5°$ and it is decreasing by about $0.00013°$ per year. Obliquity controls the seasonal variations. As the axial tilt decreases the differences between the seasons decrease, but it does not alter the total amount of solar radiation received by the Earth.

ECCENTRICITY

The Earth's orbit about the Sun is not a perfect circle, but an ellipse. **Eccentricity** is a measure of the deviation from a circular orbit. A circle has an eccentricity of zero. The Sun is at one focus of the Earth's eccentric orbit. The Earth's orbit has an eccentricity which varies between zero and 0.06, but it does not vary in an orderly manner. The extreme values of any one oscillation differ from the next. The period of oscillation is also variable, averaging 96 600 years. Perihelion is the point of closest approach to the Sun and currently occurs in the northern hemisphere winter: 2–3 January. Aphelion is the point farthest from the Sun and currently occurs in northern hemisphere summer: 5–6 July. The nearer the Earth is to the Sun the faster it moves, so northern hemisphere winter is shorter than the summer, by about 7 days at the present time. The reverse is true in the southern hemisphere. Therefore the eccentricity has the effect of varying the difference in the amount of solar radiation between aphelion and perihelion. At present with an eccentricity of 0.0017, the difference is 7 per cent. The solar radiation received in January is 3.5 per cent stronger more than average and correspondingly less in July. Northern hemisphere winters are not warmer (and summers cooler) than

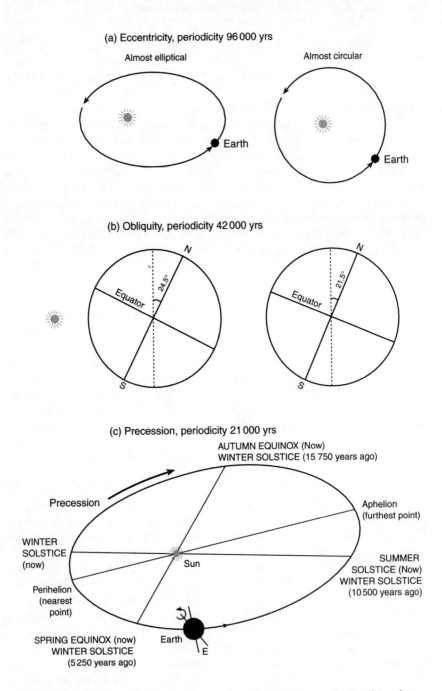

Figure 3.5 The Milankovitch cycles illustrating the variations in the Earth's orbit affecting solar radiation receipt: (a) eccentricity (b) axial tilt and (c) precession of the equinoxes
Source: After Whyte (1995)

their southern hemisphere counterparts because the difference in land mass between the two hemisphere balances the differences in radiation receipt. When the orbit is most elliptical the seasonal range in the solar beam reaching the Earth is about 25 per cent between aphelion and perihelion (Rogers, 1993).

PRECESSION

The Earth is not a perfect sphere but an oblate spheroid which means it bulges at the equator. Due to gravitational interaction with the other planets, particularly Jupiter, there is a wobble or gyration of the Earth's axis of rotation. Currently the North pole star, the star to which the Earth's axis points in the northern hemisphere, is Polaris. At night it remains fixed while the other stars appear to rotate around it. The wobble means that the pole star changes. In just over 10 500 years the pole star will be Vega. The positions of perihelion and aphelion also move in space thus shifting the location of the Earth's orbital ellipse. This process is called **precession**. Precession changes the position of the equinoxes (the points where the ecliptic and celestial equator cross). This has the effect of changing the length of the seasons by changing the time at which the Earth is closest and furthest from the Sun. For example 10 500 years ago January occurred at aphelion, therefore winters in the northern hemisphere were longer than the southern hemisphere. Two periodicities are apparent at 23 000 and 18 800 years. However, the former has the strongest signal. Changes in the position of the equinoxes alter the timing and distribution of solar radiation received but do not affect the total amount.

EFFECT OF ORBITAL CHANGES

Adhémar (1842) considered that it was the precession of the equinoxes that caused ice ages. The hemisphere which had the longest winter would experience an ice age. However, it was pointed out that the total amount of solar radiation received in a year was not affected so this in itself could not cause an ice age (Imbrie and Imbrie, 1979). Adhémar was unaware that the shape of the Earth's orbit changed over time. In 1864 Croll considered both the eccentricity of the Earth's orbit and the precession of the equinoxes (Croll, 1867). He concluded that an ice age would occur because of the affect of these two cycles on the seasonal distribution of solar radiation (Imbrie and Imbrie, 1979). A hemisphere's winter would be much longer (36 days) and colder than usual when it is inclined away from the Sun and the Earth's orbit has a high eccentricity. With less sunlight temperatures would drop and snow would be able to accumulate. This would result in a positive feedback effect with the snow reflecting more solar radiation enhancing the cooling, leading to an ice age. As the precessional and eccentricity cycles work together to create an ice age, the relative positions between them would determine the time between ice ages. When the Earth's orbit has a high eccentricity, the two hemispheres would alternately be at aphelion during winter due to the precession of the equinoxes. Thus an ice age would occur in one hemisphere during a 10 500-year span (half the precessional cycle) and then in the other hemisphere in the next 10 500 years. As the Earth's orbit became less elliptical the frequency of ice ages would decrease occurring only every 80 000 years between. Later work by Croll also considered the tilt of the Earth but the length of this cycle was not yet known. During the 1890s increasing geological evidence showed that the last ice age had not peaked 80 000 years ago as predicted by Croll but much later. Therefore the orbital theory of ice ages was abandoned once again.

Milankovitch began to study astronomical theories of ice ages between 1912 and 1914. As a mathematician he was able to calculate the changes in solar radiation distribution from all three orbital cycles. With the help of Köppen, a German climatologist, Milankovitch was able to determine that it was the reduction of solar radiation in summer at temperate latitudes that was important not that at the poles in winter. In 1930 he published his definitive work on climate change and the orbital variations of the Earth. He showed that at higher latitudes the radiation received was affected most by the 41 000-year axial tilt cycle whereas at lower latitudes close to the equator it was the precession of the equinoxes (22 000 years) that was important (Imbrie and Imbrie, 1979). The work of German geologists, Penck and Brückner, on glacial deposits of gravel in Alpine river beds suggested four Pleistocene glaciations. These corresponded with Milankovitch's times for the ice ages and the astronomical theory was generally accepted. Not everyone was convinced, Schaefer's work on the Alps called into question the chronology of the ice ages as proposed by Penck and Brückner and thus Milankovitch's theory. Then the new technique of radiocarbon dating allowed an absolute date to be given to the greatest glacial advance; this was put at 18 000 years ago. The Milankovitch theory predicted solar radiation to be at a minimum 25 000 years ago. Initially the discrepancy was put down to the response time required of the ice age. Increasingly though, new work pulled holes in the Milankovitch theory and by the mid-1950s it had been abandoned and new theories were put forward.

The astronomical theory was rescued by work on deep sea cores from the Indian Ocean. Hays *et al.* (1976) looked at the oxygen isotope signal contained within these cores. These cores extended back well over 500 000 years, far beyond the scope of radiocarbon dating. In addition, there were not enough other types of radioactive material to use radioactive dating. This meant they had to use other methods for dating the core such as magnetic polarity switching. Finally from the isotopic data Hays *et al.* (1976) had to tease out the frequency of the ice age cycles. This was done by a statistical technique called spectral analysis. There are no prizes for guessing that these revealed the orbital variations as predicted by Milankovitch. The variations in the O^{18} clearly showed a 100 000-year periodicity which correlates well with lower sea levels, lower snow lines and less carbon dioxide gas trapped in the ice (see Fig. 3.6).

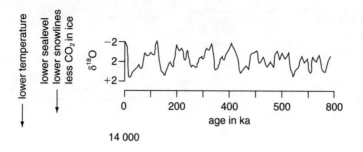

Figure 3.6 Variations in the oxygen 18/16 ratio over the past 800 000 years: there would appear to be an approximate 100 000 year periodicity
Source: Rogers (1993)

That is not the end of the story, though; in 1988 a group of terrestrial scientists (Winograd *et al.*, 1988) looked at the O^{18} record in calcite deposited from ground waters on the walls of an open water filled vault called Devil's Hole in Nevada. In this case it is the air temperature determining the O^{16} to O^{18} ratio. Unlike the ocean cores, these cores contained sufficient radioactive uranium (Uranium decays to thorium) to date the cores in the normal way. Although they came up with a similar distribution to the ocean cores, they found different dates and concluded that the **Milankovitch cycles** were not responsible for ice ages. This re-opened the debate, with both sides accusing the others of getting the date wrong. The terrestrial scientists disliked the statistical analysis of the ocean core records and the oceanographers disliked the assumption that air temperatures have the same link with global climate and the marine reservoir. Oceanographers studying coral, which can be dated using uranium, have found dates that support the Milankovitch cycles (Kerr, 1990).

Any mechanism put forward to drive the glacial cycle must be able to reproduce the nature, timing and extent of the climatic shifts that accompany the ice build-up and retreat, that are shown in the geological evidence. Over the past 800 000 years the global ice volume has peaked every 100 000 years. This coincides with the eccentricity variation. Small increases and decreases in the global ice volume have occurred at 23 000 and 41 000 years which correspond to the precession and tilt cycles. Current evidence shows that the Milankovitch cycles correspond to the timing of glacial maximum (Emiliani, 1993). Figure 3.7 shows how these three cycles may have affected the Earth's temperature. The seasonal and regional distribution of the solar radiation is the key forcing. The eccentricity variation, the only orbital change to alter the amount of radiation received, appears to drive the glacial cycles with the other orbital variations causing smaller changes in the ice volume (Whyte, 1995). There are several problems, however, with this explanation. First, the 100 000 year change in eccentricity is not entirely regular, oscillating between 95 000 years and 125 000 years, whereas the climate record shows regular

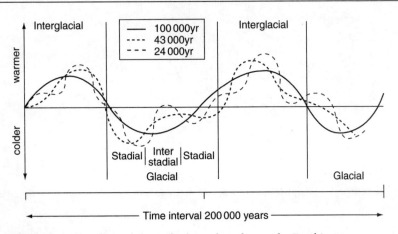

Figure 3.7 Possible effect of the Milankovitch cycles on the Earth's mean temperature
Source: Whyte (1995)

100 000-year oscillations. To explain this discrepancy it has been suggested that the climate record is not detailed enough to show these small changes (Henderson, 1998). Second, the astronomical cycles rise and fall smoothly whereas the ice record has a distinct sawtooth shape to its growth and retreat. The ice grows gradually for 80 000 to 90 000 years then disappears in a few thousand, in a period of strengthening northern summers. The third and most difficult remaining problem is that the 100 000-year cycle is the weakest of all the orbital changes but seems to have the greatest impact on Earth's climate. Therefore, although the Milankovitch cycles have been used to explain glacial epochs, the exact mechanisms for precipitating an ice age are far from understood. A major question is, can such small changes in the solar radiation alone be enough to bring about an ice age in itself or do the orbital changes trigger feedback effects within the climate system which amplify the original forcing?

AMPLIFYING THE ORBITAL CHANGES

There are several theories for how such small changes in solar radiation receipt might be amplified into a large-scale climate change. Recent modelling results suggest that the ice sheets play an important role in amplifying or driving significant variations in the climate once the Milankovitch cycles have caused large changes in the ice volume (Clark *et al.*, 1995). One suggestion is that seasonality changes act directly on the ice sheet. A reduction in summer sunshine allows the ice to build up and vice versa. Peltier and Hyde (1984) have put forward a model whereby it takes 100 000 years for the ice sheet to reach a critical size, then the ductile rock beneath the Earth's crust flows rapidly allowing the crust to sink. The ice sheet will drop and at lower elevations it will warm and with the next period of strong northern summers will melt rapidly. Broecker and Denton (1990) point out that this assumes that northern hemisphere seasonality drives the glacial cycle and yet evidence shows that glaciers grew and retreated in the southern hemisphere at the same time. For the glacial cycles to be driven only by orbital changes affecting the northern ice sheet, then the seasonality changes in northern hemisphere would have to be very large to produce so quickly changes in the southern hemisphere. Although mechanisms have been proposed such as sea level change or increased solar reflectivity from northern ice sheets, these do not solve all the problems.

The last glacial maximum occurred around 17 000 BP (years before present); this was when the global ice volume was at its maximum and ocean temperatures at their lowest. The average global surface temperature was 8°C lower than present-day values but some regions may have had temperatures 20°C lower than today. Britain experienced a mean annual temperature of around −10°C (Whyte, 1995). The glacial maximum occurs about 6000 years after the minimum solar radiation values and appears to be the time necessary for the global climate to respond to changes in solar energy. At around 13 000 BP solar radiation values recovered to their maximum. The warming continued until about 11 000 to 10 000 BP when a sudden cooling occurred in which glaciers advanced. In the North Atlantic Ocean and northern Europe temperatures decreased and remained low for some 700 years. This event is called the **Younger Dryas** cold phase. It ended very abruptly with a 7°C warming within a 50-year period (Dansgaard *et al.*, 1989). The whole change from retreat to advance to retreat may have occurred over less than 400 years according to faunal data. However, some ice core records suggest changes of 10 times faster. These changes are too abrupt to be related to variations in solar radiation receipt and instead have been associated to variations in the thermohaline deep circulation (oceanic conveyor belt) (Lehman and Keigwin, 1992). Modelling experiments have suggested that as the northern hemisphere recovered from the last glacial maximum a sudden influx of fresh water from the melting of the Laurentide Ice Sheet shut down the Atlantic conveyor belt. This led to an abrupt return to the ice age for the North Atlantic (Rahmstorf, 1997).

Broecker and Denton (1990) propose that the increase in ice sheets is a symptom of a much wider range of climatic changes. Changes in rainfall and evaporation patterns led the deep water oceanic circulation to flip from one state to another very different one. This meant that the atmospheric and oceanic general circulation was disrupted. Heat was no longer distributed from the equator to the poles in the same way. This led to a change in climate. The ice sheets then grew or shrank in response to these changes. Ice cores from Greenland show 20 per cent less carbon dioxide in the atmosphere during the Younger Dryas. Broecker and Denton (1990) conclude that the oceans, which act as a major sink of carbon dioxide, must have taken up this extra. This was achieved by an increase in efficiency of the biological pump, the phytoplankton, which takes up the carbon dioxide. If oceanic circulation is slowed, the phytoplankton stay on the surface longer, enabling them to take in more carbon dioxide. Evidence for changed oceanic circulation comes from microfossils and foraminifera shells. Broecker and Denton (1990) suggest that as the glaciers receded in North America, there was a build-up of fresh water. This water was then released into the North Atlantic. This influx of fresh water stopped the normal thermohaline circulation. The North Atlantic part of the conveyor belt which takes heat from the tropics to high latitudes became sluggish and slowed almost to a stop. Such oceanic circulation changes would then affect the atmospheric circulation. The cooling arising from the reduction in circulation allows salt levels to build up again to renew deep water formation.

This theory can now be applied to glacial cycles with evidence again coming from ice cores and the paeleotemperatures derived from them. Ice cores show that Greenland and Antarctica in the last ice age were $10^{\circ}C$ colder than today and they warmed in step. Also that the carbon dioxide concentration in bubbles of ancient air frozen in the ice is two thirds of the interglacial level. These features could all be explained by a sluggish or stationary ocean conveyor belt. However, it still relies on the weakest of the Milankovitch cycles, and small changes in seasonality, to force the change and stop the thermohaline circulation (Broecker and Denton, 1990). It could be that a small change may slow the circulation, the slower ocean will take up more carbon dioxide cooling the Earth further. There are other factors which could contribute to the overall cooling. Methane, which is also a greenhouse gas, was half interglacial levels. Dust levels were higher. The more dust, the higher the albedo and less solar radiation received to warm the Earth. Even allowing for all these changes does not account for the entire temperature drop that occurs during a glacial maximum. Broecker and Denton (1990) suggest that changes in cloud characteristics, increasing the albedo of the Earth, could provide the necessary temperature decrease. There are critics of this work who, looking at microfossil data, say there is evidence of a continuing circulation (see Zahn, 1992, for overview).

Henderson (1998) reviews other recent suggestions for the forcing mechanism of the quaternary glacial cycles. Muller and MacDonald (1995) proposed that maybe the ice ages were due to the 100 000 year wobble of the plane of the Earth's orbit. This would not appear to affect radiation receipt. However, it was suggested that the wobble moved the Earth into an area of enhanced cosmic dust, reducing solar radiation, but no evidence has been found for this. However, renewed interest in the subject meant that there is increasing evidence to suggest that the North Atlantic lags behind the southern ocean. It may well be the southern hemisphere oceans that control the glacial cycle not the north.

Other work focuses on the reduction in the amount of carbon dioxide in the atmosphere during the last glacial maximum. It has been suggested that iron may act as a fertilizer for oceanic biological activity. With more iron in the sea biological activity would increase, leading to less carbon dioxide in the atmosphere. The theory relies on orbital changes causing small local climate changes which result in the southern oceans taking up iron and then carbon dioxide. Then the oceanic and atmospheric changes amplify the original small change into a global cycle. Pollock (1997) has suggested that silicate enrichment of the waters upwelling off Antarctica could lead to decreased levels of atmospheric carbon dioxide without recourse to the Milankovitch cycles. Carbon dioxide levels would be reduced through changes in both biological activity and the solubility of carbon dioxide.

Ever since the discovery of ice ages the most acceptable theory put forward is that variations in the Earth's orbit so change the radiation distribution that an ice age ensues. The exact translation, however, from orbital changes into climate changes is still under debate. It could be that the Milankovitch cycles are merely a subtle twist in a complex chain of events. With regards to global warming a disturbing feature of the glacial cycles is the rapidity of the change in climate in some instances. During the glacial/inter–glacial cycles mean global surface temperatures may have fluctuated by around 5–7°C. At mid to high latitudes this variation could be as much as 10–15°C. It was thought that the onset of glaciations was very gradual but it could be as rapid as 1°C per century and figures as much as 5°C per decade have been suggested for some of the changes that have occurred.

Summary

All the mechanisms for climate change introduced in this chapter operate on too long a time-scale to be responsible for global warming. However, it emphasizes the need to appreciate time-scales and to recognize that the forcing must be able to reproduce the climate change. This chapter has described the ways in which the climate record can be reconstructed and dated. The scale of temperature changes and time taken for these changes to occur provide a base line when coming to assess the effects of global warming. Current understanding of the geological climate suggests that small radiative forcings can be amplified by the climate system itself to produce large-scale global climate change and that this can occur over relatively short time periods.

References

Adhémar, J.A. 1842: *Révolutions de la mer*. Paris: Carilan-Goeury et V. Dalmont.

Appel, P.W.U. and Moorbath, S. 1999: Exploring Earth's oldest geological record in Greenland. *Eos* 80, 257, 261, 264.

Broecker, W. and Denton, G.H. 1990: What drives glacial cycles. *Scientific American* 262, 43–50.

Budyko, M.I. 1986: *Evolution of the biosphere*. English translation. Dordrecht: D. Reidel.

Charlson, R.J., Lovelock, J.E., Andreae, M.O. and Warren, S.G. 1987: Oceanic phytoplankton, atmospheric sulphur, cloud albedo and climate. *Nature* 326, 655–61.

Clark, P.U., MacAyeal, D.R., Andrews, J.T. and Bartlein, P.J. 1995: Ice sheets play an important role in climate change. *Eos* 76, 265.

Condie, K.C. and Sloan, R.E. 1998: *Origin and evolution of Earth: principles of historical geology*. London: Prentice-Hall.

Croll, J. 1867: On the eccentricity of the Earth's orbit and its physical relations to the glacial epoch. *Philosophical Magazine* 33, 119–31.

Croswell, K. 1994: Life and times of a star. Inside Science no. 76. *New Scientist* 144 (no. 1953).

Crowley, T.J., Baum, S.K. and Kim, K.Y. 1993: General circulation model sensitivity experiments with pole–centre supercontinents, *Journal of Geophysical Research-Atmosphere* 98(D5), 8793–800.

Dansgaard, W., White, J.W.C. and Johnsen, S.J. 1989: The abrupt termination of the Younger Dryas climate event. *Nature* 339, 532–4.

Dawson, A.G. 1992: *Ice age Earth: late quaternary geology and climate*. London: Routledge.

Emiliani, C. 1993: Milankovitch theory verified. *Nature* 364, 583–4.

Gore, A. 1992: *Earth in the balance: forging a new common purpose*. London: Earthscan.

Greenberg, R. 1989: Planetary accretion. In Atrey, S.K., Pollack, J.B. and Matthews, M.S. (eds), *Origin and evolution of planetary and satellite atmospheres*. Tucson: The University of Arizona Press.

Hays, J.D., Imbrie, J. and Shackleton, N.J. 1976: Variations in the Earth's orbit: pacemaker of the ice ages. *Science* 194, 1121–32.

Henderson, G. 1998: Deep freeze. *New Scientist* 157 (no. 2121), 28–32.

Henderson-Sellers, A. 1983: *The origin and evolution of planetary atmospheres*. Bristol: Adam Hilger.

Houghton, J. 1994: *Global warming: the complete briefing*. Oxford: Lion Publishing.

Huggett, R.J. 1991: *Climate, Earth processes and Earth history*. Berlin: Springer-Verlag.

Hunt, L. 1998: Send in the clouds. *New Scientist* 158 (no. 2136), 28–33.

Imbrie, J. and Imbrie, K.P. 1979: *Ice ages: solving the mystery*. London: Macmillan Press.

Kasting, J.F. and Toon, O.B. 1989: Climate evolution. In Atrey, S.K., Pollack, J.B. and Matthews, M.S. (eds), *Origin and evolution of planetary and satellite atmospheres*. Tucson: The University of Arizona Press.

Kerr, R.A. 1990: Marking the ice ages in coral instead of mud. *Science* 248, 31–3.

Kuhn, W.R., Walker, J.C.G. and Marshall, H.G. 1989: The effect on Earth's surface temperature from variations in rotation rate, continent formation, solar luminosity and carbon dioxide. *Journal of Geophysical Research* 94, 11,129–11,136.

Lehman S.J. and Keigwin, L.D. 1992: Sudden changes in North Atlantic circulation during the last glaciation. *Nature* 356, 757–62.

Lovelock, J. 1988: *The ages of Gaia*. Oxford: Oxford University Press.

McElroy, M.B., Kong, T.Y. and Yung, Y.L. 1977: Photochemistry and the evolution of Mars' atmosphere: a Viking perspective. *Journal of Geophysical Research* **82**, 4379–88.

Morton, O. 1998: Flatlands. *New Scientist* **159** (no. 2143), 36–9.

Muller R.A. and MacDonald, G.J. 1995: Glacial cycles and orbital inclination, *Nature* **377**, 107–8.

Pearce, F. 1989: *Turning up the heat*. London: The Bodley Head.

Peltier, W.R. and Hyde, W. 1984: A model of the ice age cycle. In Berger, A., Imbrie, J., Hays, J. Kukla, G. and Saltzman, B. (eds), *Milankovitch and climate: understanding the response to astronomical forcing*. Nato ISI Series C, Volume 126. Dordrecht: Reidel.

Pollack, J.B. 1971: A nongrey calculation of the runaway greenhouse: implications for Venus' past and future. *Icarus* **14**, 295–306.

Pollack, J.B. 1979: Climatic change on the terrestrial planets. *Icarus* **37**, 479–553.

Pollack, J.B. 1990: Atmospheres of the terrestrial planets. In Beatty, J.K. and Chaikin, A. (eds), *The new solar system*. Third edition. Cambridge: Cambridge University Press.

Pollock, D.E. 1997: The role of diatoms, dissolved silicate and Antarctic glaciation in glacial/interglacial climatic change: a hypothesis. *Global and Planetary Change* **14**, 113–25.

Portney, K.E. 1992: *Controversial issues in environmental policy: science vs economics vs. politics*. Newbury Park: Sage.

Rahmstorf, S. 1997: Ice-cold in Paris. *New Scientist* **153** (no. 2068), 26–30.

Rogers, J.J.W. 1993: *A history of the Earth*. Cambridge: Cambridge University Press

Schneider, S.H. and Londer, R. 1984: *The coevolution of climate and life*. San Francisco: Sierra Club Books.

Walker, J.C.G., Hays, P.B. and Kasting, J.K. 1981: A negative feedback mechanism for long-term stabilization of Earth's surface temperature. *Journal of Geophysical Research* **86**, 9776–82.

Watson, A.J. and Lovelock, J.E. 1983: Biological homeostasis of the global environment: the parable of Daisyworld. *Tellus* **35**, 284–9.

Wayne, R.P. 1992: Atmospheric chemistry: the evolution of our atmosphere. *Journal of Photochemistry and Photobiology* **62**, 379–96.

Whyte, I.D. 1995: *Climatic change and human society*. London: Arnold.

Winograd, I.J., Szabo, B.J., Coplen, T.B. and Riggs, A.C. 1988: A 250 000 year climatic record from great basin vein calcite: implications for Milankovitch theory. *Science* **242**, 1275–80.

Zahn, R. 1992: Deep ocean circulation puzzle. *Nature* **356**, 745–6.

4

Historical climate change

Introduction

The previous chapter looked at very long-term climatic changes occurring over tens of thousands of years. What relevance does this have to us who live but for a fraction of that time? If global warming occurs on those sorts of time-scales how can it possibly affect us? Global warming, however, is predicted to occur on short time-scales, over decades to centuries. If we are to assess the impact of global warming on the climate we also need to consider natural variations in climate on these time-scales. Can natural climatic changes occur on short time-scales and what is the range of that change? What causes climate changes on these shorter time-scales? Are the changes large enough to be perceptible, and can these natural climate changes affect the world around us and the way we live?

Evidence for climate change over historical time periods

The evidence for short-term climatic variation, from 1 to 100 years requires data from half a millennium (500 years). The sources of such evidence can be split into two: instrumental data and proxy data such as historical documentation, ice cores and dendrochronology. In preparing a climatic record from these data sets the same sorts of problems present themselves as with the geological climate record. The same types of questions need to be asked of the data. What type of data are available, for example, temperature, wind speed, precipitation etc.? How reliable are the data? Are data available for the whole globe or just a localized area? Can an exact date be applied to the data? How long is the time series and what is the resolution? Is there a lag time between when the climate record was laid down and the observation? In our 'information age' it is sometimes hard to appreciate the difficulties in preparing a climatic data base for such a recent part of Earth's history. The following is a synopsis of some of the main ways in which the historical climate can be reconstructed. Comprehensive reviews of the various techniques can be found in Bradley and Jones (1992), Hecht (1985) and Wigley *et al.* (1981).

Meteorological instrument records

It seems obvious that we should regard instrumental data as the most reliable and easily available. Nowadays a large number of different meteorological parameters, such as surface pressure, rainfall, temperature, relative humidity, wind speed and direction, are measured by a network of professional observers using sophisticated reliable equipment at specific observing sites. That was not the case a few hundred years ago. The earliest instrumentation to appear was the rain gauge. This was used in Korea as early as the 1400s but no records of the early measurements survive. No one is quite sure who invented the thermometer but it is attributed by some to Galileo in 1590. The lack of a universal measurement scale and poor reliability limit the use of early thermometer readings. The mercury pressure gauge or barometer was invented by Toricelli (1608–1647) in 1643. Dependable instruments have been around since about 1700. In Europe some measurements have been taken since the 1600s, but for most places only a few decades of data are available. Where instrumental records do exist, they rarely go back beyond 1750 and this means that the instrumental record covers only half the time span needed for research into historical climate change. Even reliable instruments do not make for reliable observations. With no professional network early observations were made by enthusiastic amateurs. In Great Britain these were usually men with independent incomes, and consequently time to devote to a hobby, such as the local squire or clergy. Early thermometers and rain gauges were often exposed in the wrong sites. Different units and scales were used. In the late 1700s to early 1800s 77 different temperature scales were being used. Dates and times are usually recorded with the observations but even time did not become standardized until the 1800s. In Great Britain it was the advent of the public railway system that required each place in the country to maintain the same time. The observations reflect the state of the atmosphere at that time but may only have been taken once a day. Night-time observations are often missing.

From this hotch potch of data climatologists try to reconstruct records of the past climate. This involves patching together data from various different observers and sites. Overlapping data sets can be analysed and made consistent with one another in order to obtain a long data record. When reconstructing a data set, the scales used, the observing technique and the exposure site all have to be considered. Surface temperature and precipitation are the most common variables available for this type of reconstruction (Hulme and

Jones, 1994). There are several surface temperature data sets available for Great Britain, the earliest one dating from 1659 for central England, closely followed by a data set for Edinburgh, Scotland (Duncan, 1991 and Manley, 1974). There are many problems in trying to reconstruct a hemispheric temperature sequence, not least the lack of observations available for the oceans. So far only surface observed data on land has been discussed. Ship-borne instruments do not become available until about 1850. Organized meteorological services with standardized meteorological equipment and trained observers appeared in the later half of the nineteenth century. For example the UK Meteorological Office was created in 1854 and the USA weather bureau in 1870. The World Meteorological Organisation was formed in 1873. Upper air data, which measure various meteorological variables with height have only become common since the Second World War.

All the data discussed so far are called *in situ* measurements, that is the variable is measured directly at the location. Remotely sensed data occur when an instrument measures a variable in a different location to the instrument. This type of data has only become widespread since the advent of satellites in the 1960s. So unfortunately satellite climate records are extremely short. Satellites remotely sense the Earth's atmosphere and surface providing all types of meteorological variables. Data for climatological studies are typically provided globally every three hours. This may be a climatologist's dream come true but the sheer volume of data that this provides can be a problem in itself and this is not the only problem with satellite data. Satellites have a limited lifetime and need to be replaced. The replacement inevitably reflects the technological advances that have been made and so does the onboard instrumentation. Each satellite change brings an instrument change and a potential discontinuity in the data set. To compile an error-free data set an instrument first requires calibration. This means that we need to know what an instrument reading of 10 units corresponds to, for example, if the instrument measures temperature does 10 units mean 10°C or 9.8°C? The wrong calibration can lead to erroneous measurements. Although the data can often be corrected it may take some time before it is noticed. Some meteorological variables are just plain hard or impossible to measure from satellites at the present time. One of these is soil moisture content which measures just how dry or wet a soil is. This is very important to agriculture and consequently it would be useful to measure it accurately and advise farmers on the present and future climate. Therefore although remotely sensed data from satellites provide some of the most objective and complete data sets available, there are limitations for climate change studies.

Proxy data

Proxy data reveal climatic data through the interpretation of some form of related data. For most of the techniques described in this section the essence

of that interpretation is quite straightforward. In practice, however, interpreting proxy data requires complex techniques allowing the data to be standardized, that is removing extraneous variations in the data record, and
calibrated, thus relating the data to some baseline. The following is therefore
a simplified account of the techniques used.

HISTORICAL DOCUMENTATION

Under this heading is included written data that are not based on instrumental readings but other more anecdotal information. This type of information
can be found in hundreds of different written texts. It may be direct information on the weather as in diaries and journals, or indirect evidence, such as
crop yields, harvest dates and grain record prices, which will reflect the
weather of the time. The data have to be turned from these vague descriptions
into some form of numerical description of the weather, so they are a type of
proxy data. Obviously there is a great deal of uncertainty in this process
which decreases the reliability of the data. For example in the case of grain
yield good weather will lead to higher yields and so to lower prices. However,
what do we mean by good weather and how would changes in agricultural
practice at the time affect yields and prices? What about other social and economic developments that may affect prices? Another problem with such subjective writings is the motive of writer. It may be a diarist with no other
motive than to provide an account of the daily goings on in which the weather
is a mere adjunct. Or the writer could be making a report as to why an expedition to find new riches failed, in which case the exaggerated weather conditions just might keep the writers head intact, literally. Where possible it is
best to compare documentation with other reports and evidence. There is a
large amount of material out there, but to pull that information together is
very time-consuming. Calibration is usually achieved by attempting to overlap part of the historical documentation with early instrumental data. Some
sort of mathematical relationship can then be derived to link the two data
sources and thus provide a way of interpreting earlier documentation.

   Of course the date will usually appear on the documentation but accurate
dating can be a greater problem than you may think. Different countries
started their new year on different dates. Then there was the change from the
Julian calendar, with no fixed new year's day, to the Gregorian calendar which
begins, as now, on 1 January. Apart from the change to the start of the year
other alterations to the occurrence of leap years were made. These revisions
to the calendar by Pope Gregory XIII happened in most of Europe in 1582 but
the Russians changed as late as 1917 after the Bolshevik revolution. Rather
than stating the year, some documents refer to being in the X year of the reign
of monarch Y. Very often the same weather event is attributed to different
years, but sometimes extreme weather can occur in a series of years and it is
a mistake to lump them together as one event. So, obtaining the correct year
requires careful consideration. Probably the best hoped for time resolution is

days, though in ships' logs this might be greater. Annals and account books may describe the monthly or seasonal weather, for example if it was a particularly wet or dry July. This type of data requires that the civilization had developed writing as a means of recording events. Extreme weather events were recorded by the ancient Babylonians on inscribed stones. However, most material comes from Europe, China and India. The information usually refers to the local climate and gives little insight into the global climate of the time.

SEDIMENTARY DATA

Ice and ocean cores which have already been discussed in Chapter 3 are also of use in studying historical climate changes. Similarly, varves can be used for historical climate reconstruction as they provide information on rainfall and streamflow. Rivers also provide proxy data through flood records. The variables provided are rainfall and evaporation from flood and lake levels. The length of the time series for some of these records are impressive. Lake levels also provide information on the climate. For Lake Saki in the Crimea the thickness of the yearly mud levels is dependent on the summer rainfall. The length of the climatic record is long and can be detailed back to 2300 BC. Longer time series exist for beds and lakes formed around the margins of ice sheets during post-glacial melting. This has been achieved by using short-lived but overlapping lakes to form one long composite time series. However, the technique is not without difficulties. For the Nile, some flood records go back to 3100 BC, although they mostly date from 622 AD. Flooding in the Nile delta depends on summer monsoons over Ethiopia. The data are really only useful for information about an individual year not a shorter time period. With varves, days or weeks can be ascertained but for lake levels the resolution may be as large as 15 years. Also the data are scattered around the globe from individual lakes and rivers. Reliability of the data also poses a problem. In the Nile prolonged silting changes the zero level making it difficult to ascertain the flood levels.

FAUNA AND FLORA

There are several important ways in which flora and fauna can help with the climatic record. The most well known are the analysis of pollen grains and tree ring analysis called dendrochronology. Pollen grains allow both temperature and rainfall to be assessed and have various features that make them ideal for providing proxy data. Plants produce pollen in profusion and they are extremely resistant to decay. It is possible to identify a plant species from its pollen grain and this will provide information about the climatic conditions of the area. As each pollen grain can be counted, the amount and variety of each species in the locale can be estimated. This rests on the assumption that the pollen grain sample is representative of the plants growing in that area. In some locations, such as peat bogs, the pollen grains are laid down in layers with the peat. Thus changes in the climate can be reconstructed from recording the changes in vegetation. Finally pollen grains can be dated exactly by radiocarbon dating.

Although there is a potential to cover the globe with this type of analysis, in practice it is limited to only a few sites. These sites provide local coverage on a spatial scale but several of the sites provide data stretching back 125 000 years or more. The record length is dependent on site conditions and dating. Unless the pollen is in peat or lake bed sediments it can be difficult to date within 100 years. Furthermore radiocarbon dating is expensive and is limited to 70 000 years. The first step in reconstructing a past climate from pollen grains is to determine a species present-day response to climatic indicators, such as temperature and precipitation. This present-day relationship is then used to interpret the past fossil pollen record in terms of the past climate. Problems in reliability lie with the assumption that each forest type corresponds to the same sort of climate as today. Although there is reason to believe that vegetation is fairly responsive to climate, there are known to be some lags. In Europe there are long delays between the establishment of a climate regime and certain trees arriving. It can take centuries for these trees to become dominant. These delays have been detected by the change in insect population which has a quicker response time, particularly some beetle species. In contrast in North America there is little time lag between the climate change and tree response. Human activities have also affected plant species and bog development which again will alter the pollen record.

Dendrochronology allows the determination of temperature and rainfall by looking at the widths of tree rings which have been found to vary in response to changes in rainfall and temperature. Tree rings are formed by many trees. There is a growth spurt in spring when thin-walled cells are formed. The growth then slows down in autumn when thick walled cells are produced. The difference between the two types of cells produces the distinct boundary: the ring. Age is determined by counting the number of rings from the trunk of a felled tree. If the tree was still growing when felled then in the ring widths you have a climatic record from the day it fell, back to when it first grew. In the case of the long-lived Bristlecombe pine this can allow the climate record to go back to 343 BC. The record length that can be obtained, however, is not limited by the tree's life. Although there are small variations in tree ring width, due to local factors, over quite a large area a similar pattern in the rings emerges which is attributable to the climate of that region. The distinctive pattern of the rings then allows that record to be overlapped with those from trees felled earlier, and used for building timber or that have been fossilized or preserved in peat bogs, forming one long record. The determination of the climate from ring widths depends on the tree's response to climate. This is not as straightforward as the description above portrays and depends on the tree biology as well as the climate. Various factors due to the genetic make-up of the tree and the environment introduce variations into the ring widths. Complex techniques are used to remove these variations revealing the trend of the underlying climate variable. Generally, ring widths in trees at the upper altitude limit and at the poleward limit are related to summer temperature. Tree ring width from trees at the warm arid limit are determined more by precipitation. The ring width depends on the previous 15 months weather and so the record has a resolution of about one year. Tree rings are particularly good at showing year to year and decadal variations in the climate (Pearce, 1996).

The recent climatic past

Having established the various ways in which climatologists can determine the relatively recent past climate, let us briefly take a look at how that climate has altered. The evidence for historical climate change comes not just from reconstructing the climate record but from a diverse set of disciplines from archaeology to geology, history to botany. There are many reviews of climate history (Bradley and Jones, 1992; Hecht, 1985; Wigley *et al.*, 1981) and particularly interesting are those by the late Hubert Lamb (Lamb, 1982) one of the most influential workers in this field. This overview draws on these works and starts at 5000 BC (7000 BP), the period that marks the dawn of civilization. In this some of the major political, economic and societal upheavals are indicated. These provide signposts for the climate history. It is not intended at this stage to contemplate whether the climate was the cause of these effects, that will be discussed at the end of the chapter.

Early history

Between 5000 BC and 3000 BC the global average temperature was probably some 1 to 3°C warmer than it is now. From 4000 to 2440 BC it is possible that the African Monsoon occurred as far north as the Sahara, implying greater vegetation and a more moist regime in this area. Evidence of this comes from rock paintings which depict agriculture and animals such as hippopotamus in an area which now sees very little rainfall. During this period the Early Kingdom cultures in Egypt flourished next to the Nile which may have discharged three times the amount it does now. From 3500 to 3000 BC there was a cool period called the Piora oscillation during which glaciers advanced. This probably lasted no more than 4 centuries preceding an unsettled but, cooler and drier climate which set in from 3000 to 1000 BC. Proxy evidence for this comes from tree ring widths of the Bristlecombe Pine in California, glacier advance and retreat and the upper tree line. The great civilizations of the Sumers on the Tigres–Euphrates, the Egyptians on the Nile, and the Indus civilization in the Indus Valley in India all appeared and were at their height during this period. The Egyptian civilization provide accounts of the Nile river level which suggest this was lower between 2200 and 2000 BC indicating a drier regime at this time. Agriculture appeared in Europe and China. Cultivation in Dartmoor occurred up to 450m above sea level whereas now it is about 300m. Megalithic monuments appeared in Europe. Originally archaeologists thought that civilization had spread outwards from its cradle in Sumer to the Hebrides. From the late 1970s comes the view that the megalithic monuments of the Orkneys and the tombs in those early civilizations were contemporaries, and great monuments had burst forth at the same time.

From 2000 BC there were abrupt temperature changes and by 500–200 BC the average temperature in Europe was 1°C lower than that in the warmest post glacial period. Between 1000 BC and 1 BC there is evidence that northern Europe became wetter. There are also indications that central Europe experienced wet phases during this last millennium BC. At this time there was also a general migration of people from north to south. The ancient civilizations disappeared during this period but Rome was founded in 753 BC. The period between 403 and 222 BC is known as the Warring States Period in China. The following 400 years appears to be a period of stability, not just in China but also for the Greeks and America (Pearson, 1978). While Lamb (1982) concludes that between 120 BC to 400 AD there was global warming, more recent evidence confirms the classical view that during this period, while Rome was in decline, central Europe suffered a cooler, more arid climate (Brown, 1994). There was another period of migration in central Europe from east to west. From then on more data become available, so more can be said about individual regions. At low latitudes, in Antarctica, the North Pacific and parts of the Arctic the warm period continued until about 1200 AD. Japan and China remained cold apart from a slight warming between 650 and 850 AD. In Europe the Roman Empire was sacked in 476 AD. There were extreme winters with ice on the Dardanelles in 763/764 AD, the Adriatic 859/60 AD and on the Nile and Bosphorous in 1010/1011 AD. Glaciers advanced and tree rings indicate a cold period. In North America, Europe and the European Arctic this cold period was followed by a renewal of warmth in the tenth to twelfth centuries. This is commonly referred to as the Medieval warm period or Little Optimum as temperatures approached those seen in the warmest part of the post-glacial centuries. The eighth to eleventh centuries saw the main Viking raids on north-west Europe and the colonization of Iceland and Greenland. Wine production requires a mean annual temperature of 15°C and nowadays most of Britain experiences a climate unsuitable for viticulture (Spellman, 1999). However, in the Medieval Optimum vineyards flourished in England as far north as York.

From 1000 to 1600 AD there is archaeological evidence to suggest a drier regime in central and south-west America. In North America, the climatic trends appear similar to that which occurred in Europe. The spread of whaling indicates warm summers but this was to be followed by a period of pronounced cooling which was also to be found in Europe. By 1200 AD there was a climatic deterioration in the northern European countries and this spread to the rest of Europe by the 1300s. From about this point the whole of Norse society went in to decline. Europe suffered severe wind storms and sea floods. The English ports of Ravenspur and Dunwich were lost to sea storms at this time. The period 1284 to 1311 was very dry in Europe and was then followed by a series of wet years until 1321. The summers during many of the decades of the 1300s were dry and drought occurred. Very severe winters occurred in the early 1400s. There was an increase in wetness and St Anthony's Fire, a disease caused by ergot blight on the kernels of rye from a damp harvest, was rife. This exacerbated the effects of the Black Death which arrived from China sometime between 1348 and 1350. The period of 1430 to 1480 was also a period of cold in Northern Europe. Most desertions of villages in England took place at this time. The climate was beginning to enter a cool phase, cooling by as much as 1°C to 1.5°C and there is evidence to suggest that this was a global phenomena which is called the **Little Ice Age**.

The Little Ice Age

It is difficult to give an exact period for the Little Ice Age as it varied from place to place. Lamb (1982) defined it as the period between 1420 and 1850 with 1550 to 1800 being the maximum. The 1500s were the peak of the Renaissance and the Reformation. There was a commensurate increase in documentary evidence and, by the 1700s, systematic instrumental observations (Bradley and Jones, 1992). In Europe between 1500 and 1800 it is believed the mean winter temperature was 1.3°C lower than for 1880 to 1930. Temperature records for England show a mean temperature 0.9°C lower than in 1920 to 1960. In the 1690s all the seasons were extremely cold and the run of cold wet years from 1570–1600 and 1690–1700 resulted in an advance of the Alpine glaciers. The growing season in the 1690s was shortened by 5 weeks compared with the warmest decades (except the nineties) of the twentieth century, and this probably lasted for 30 to 50 years. The average temperature of England for the period 1670 to 1700 suggests that the number of days per year that snow was lying on the ground was 20–30 compared with 2–10 days today (Lamb, 1982). In central Europe temperature and precipitation reconstructions by Pfister (1992) reveal that during the Little Ice Age the climate was more variable but that winter and spring were both colder and drier compared to the twentieth century. However, there was no tendency for the summer months to be colder.

During the Little Ice Age many European rivers froze and the Thames in London was no exception. On many occasions the ice was thick enough for people and carriages to cross the river and even for frost fairs to occur (*see* Table 4.1). The old London Bridge was wooden and built on many piers. This allowed the build-up of stagnant cold water which, when cold enough, was able to freeze over. The first frost fair took place during the winter of 1607/08. The frost began mid-December and by mid-January the ice was

Table 4.1 Dates of frost fairs on the River Thames

Year	Notes
1607/08	
1615	Possible frost fair.
1683/84	The Thames was frozen from the beginning of December until approximately 5 February. The fair occurred in the latter half of January.
1688/89	The frost lasted from 20 December to 6 January.
1715/16	Severe frost from 24 November to 9 February.
1739/40	A very severe winter with the frost lasting from 24 December to 17 February
1788/89	A hard frost from 25 November to 14 January.
1813/14	Another severe winter with frost from 27 December to 5 February. The fair occurred on 31 January.

Source: Brazell (1968)

thick enough to bear the weight of people. Stows English Chronicle of 1611 describes the scene at London Bridge as one of merriment with people pursuing a variety of pastimes. London Bridge had survived the great fire of London in 1666 but in the rebuilding after the fire many of the Thames tributaries were covered over. During 1683/84 perhaps the most famous frost fair occurred. The Thames was completely frozen for two months and the ice was reported to be 11 inches thick. John Evelyn, a diarist of the time recorded many entries about this fair. The fullest entry comes on 24 January 1684: 'The Thames before London still planted with booths in formal streets ... coaches plied from Westminster to Temple ... a bull baiting, horse and coach races, puppet plays, cookes, tipling and other lewd places ...'. While all this festivity took place on the ice, inland there was real hardship 'the trees not only splitting as if lightning had struck, but men and cattle perishing in divers places ... and all sorts of fuel so dear that there were great contributions to preserve the poor alive'. Evelyn also records that this winter affected many other parts of Europe including Spain and other southerly areas (Wheatley, 1906). The last decade of the 1600s is particularly noted for the number of severe winters in England. Six of the winters had seasonal mean temperatures below 3°C (Kington, 1999). Frost fairs continued throughout the 1700s but early on in the century the climate had started to improve. The recovery was not smooth. Eight winters in the 1730s were as warm as the warmest between 1900 to 1950, yet 1725 had a summer as cold as any winter with a mean temperature over June, July and August of just 13.1°C. The Thames also changed. In 1757 London Bridge was partially rebuilt with a wide central arch but still with piers. The last frost fair was in 1813/14 with a full fair running from 2 to 6/7 February. In 1831 the old London Bridge was removed and a new one built.

There is documentary evidence that the Little Ice Age was at least as widespread in Europe. Such evidence is not always conclusive proof that a climatic change occurred. The reason the Thames froze over has as much to do with the construction of the old London Bridge, the covering of tributaries, the tidal reach and industrialization as with changes in climate. The Thames, however, was not the only river to freeze. The Ouse in York had its own frost fair. Lake Constance froze and the sea around Marseilles froze in 1595 and 1638. The cold winter of 1697/98 also affected western and central Europe (Kington, 1999). Other documentary evidence points to severe hardship with rural depopulation and decreasing life expectancy in England (Lamb, 1982). More tangible evidence comes in the form of glacier advance in Europe which is indicated in documents and drawings of glaciers. Grove (1988) describes the alpine glacier advance during the Little Ice Age as follows. Around 1300 AD there was an advance of the glaciers in the European Alps, the following retreat was probably not as complete as that as during the medieval warm period. So whether it can be thought of as a forerunner to the Little Ice Age or part of it is difficult to decide. The date of the maximum advance of glaciers is best found from the Alps, and this appears to be 1600. X-ray densitometry of tree rings show a sharp decline in summer temperatures around 1570. Northern Scandinavian glaciers appear to have advanced at the same time as the alpine glaciers, whereas in southern Scandinavia the advance came in the late seventeenth century.

In Europe the date of glacial advances can be obtained from historical records. For the rest of the world such records do not exist. Thus the times of glacial advance in the rest of the world is obtained from the dating of moraines. Dates are derived from the measurements of maximum lichen diameters and the ages of trees growing on moraines (Grove, 1988). In North America the times correspond well with those from the Alps. In South America ice cores from the Quelccya Ice Cap (Thompson *et al.*, 1986) were dated using stratigraphic features, such as annual dust layers and microparticle concentrations, conductivity and oxygen isotope ratios. The Little Ice Age stands out as a period from about 1500 to 1900. There is even evidence for this advance in the Himalayas and New Zealand. The increasing body of evidence points to the fact that the Little Ice Age was a global event with a decrease in average global surface temperatures (Grove, 1988). It would be a mistake, however, to think of the Little Ice Age as a period of continuous cold weather at the same time over the entire globe. As Jones and Bradley (1992) point out, there was great regional and temporal variability, indeed they even question whether the term Little Ice Age should be used at all. The coldest period in Europe and North America was probably the nineteenth century, but in eastern Asia it was the seventeenth century. The Little Ice Age seems to be a time of great variability, where several cold decades (up to about 30 years) were interspersed with warm decades, and timing of these periods varied from region to region.

From the 1700s to the third millennium

Lamb (1982) describes the recovery period from the Little Ice Age as erratic. However, Bradley and Jones (1992) suggest that there was always a great variation in the climate. From the eighteenth century, the amount of data from Europe begins to increase rapidly with the introduction of meteorological instrumentation. The eighteenth century in Europe was a period of warmer conditions with summer temperatures in the west of the former Soviet Union being 1°C warmer than the long-term average (Jones and Bradley, 1992). By the late eighteenth century colder conditions had returned to Europe and it was at this time that the French Revolution took place. In the late 1700s France experienced famine when drought followed a severe winter in 1785. Some 55 per cent of the poor's income went on bread alone. When the French Revolution occurred in 1788/89, this figure was 88 per cent (Lamb, 1982). There was then a short period of warmth, but the first two decades of the nineteenth century brought renewed cold for most of Europe. The decade 1810 to 1819 brought the coldest run of winters since the 1690s. The early nineteenth century appears to have been a time of extreme variability. In the 1840s there were very cold summers, but 1846 was extremely hot. Towards the end of the nineteenth century Europe had a more colder climate. Then during the second and third decades of the twentieth century there was climatic warming. The growing season in England increased by around 14 days (Lamb, 1982). There followed a cooler climate in the 1950s and 1960s which was maintained throughout the 1970s up to 1980.

Evidence from the rest of the world is far more fragmentary and relies heavily on proxy data for limited areas. In contrast to Europe, Japan and China had a calmer climate with few deviations from the norm. North America had temperature variations similar to Europe with warmer temperatures throughout the eighteenth century reverting to cooler temperatures in the nineteenth century. However, proxy data for the western USA show little evidence of any cooling during the eighteenth century. In the southern hemisphere there is even less data available. In New Zealand dendroclimatic evidence shows periods of cooling in the 1840s and 1860s. Glaciers, however, have been retreating since 1840 (Jones and Bradley, 1992). Over the last hundred years glaciers have been retreating with few exceptions. This recession was halted almost globally in a few decades, notably the 1880s, 1920s and 1960s. Since the 1980s an upward trend in the global average surface temperature has been detected.

Natural causes of climate change on short time-scales

It appears that throughout the Holocene there were advances of glaciers comparable to the Little Ice Age lasting about 100 years and spaced roughly a millennium apart. The phases are mostly globally synchronous and affect both hemispheres (Grove, 1988). Proxy data and, more recently, instrumental data all point to global climate changes, albeit with significant regional variations, on the historical time-scale. Milankovitch cycles are far too long to explain such variations. This section will consider the forcings that might cause variations on historical time scales.

External forcings

SOLAR OUTPUT

The Sun is the ultimate heat source. Therefore it is logical to conclude that changes in the solar output will affect the climate on Earth. What are the possible sources of such changes? The **solar constant** was called a constant because everyone thought solar output was stable. It is only since the 1980s that satellite measurements have provided conclusive proof that the energy reaching the Earth is far from constant. Changes are detectable from month to month, on decadal time-scales and even longer. There are many types of solar disturbance which may affect the output and can be linked to changes in the solar constant.

Sunspots are perhaps the most well-known solar disturbance. They are due to magnetic activity within the plasma of the Sun. Sunspots are visible with the naked eye when the Sun is low in the sky (please note it is not recommended that anyone looks directly at the Sun). Observations of sunspots are well documented in China with some references from the first century BC Telescopic observations began in the early 1600s. However, it was not until 1845, after 17 years of observing, that H. Schwabe discovered the sunspot cycle. He found that the number of sunspots varied with a maximum approximately every 11 years. This sunspot cycle was traced back to the 1700s by R. Wolf (*see* Fig. 4.1). He developed the idea of a **sunspot number** which is the number of sunspots multiplied by ten added to the number of individual sunspots multiplied by a constant (Hoyt and Schatten, 1997). In 1908 it was realized by G.E. Hale that in fact the sunspot cycle is a 22-year one. It was discovered that most sunspots occur in pairs or groups with two principal spots. One spot has a northerly magnetic polarity and the other southerly. In the pair there is always one spot slightly ahead of the other called the leading spot. In an 11-year cycle the leading spot of the pair (or group) has one polarity in the northern hemisphere and the opposite in the southern hemisphere. In the next 11-year cycle this polarity is reversed (Hoyt and Schatten, 1997).

There have been many attempts to link sunspots with climate changes on Earth. The idea goes back many years. In 1651 a Jesuit priest called Riccioli suggested that the temperature of the Earth decreased with an increasing number of sunspots. This is quite a logical assumption, since a black area on the Sun suggests lower temperatures. The more spots, the less energy the Sun would emit. However, in 1801 William Herschel, came to the opposite

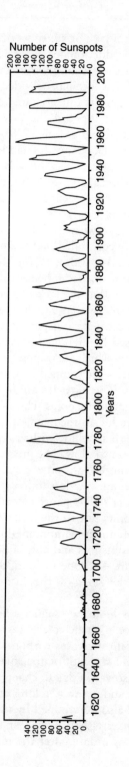

Figure 4.1 The sunspot cycle from early sixteenth century to the present day

conclusion. He noted that a lack of sunspots occurred at the same time as a lack of wheat. As wheat production depends on sunlight Herschel concluded that fewer sunspots led to less sunlight and therefore lower temperatures. Investigation of the Little Ice Age has provided some fertile ground for the study of solar variability and its link to climate change. In 1887 a German astronomer, Spörer had put forward the possibility that there were few sunspots for a prolonged period before 1716. This had been noted by an English astronomer in 1711, Denham who had written that 'there are doubtless great intervals sometimes when the Sun is free of sunspots, as between the years 1660 and 1671, 1676 and 1684'. In 1890 and 1894 an English astronomer named Maunder published the evidence for such a prolonged sunspot minimum pointing out the implications for solar–terrestrial relations. His work was largely ignored until 1976 when Jack Eddy, an American astronomer, re-examined the evidence. He concluded that a sunspot minimum had occurred at the time of the Little Ice Age (Eddy, 1976). Most of the initial evidence for Maunder and Eddy's work came from historical documentation of the time. A major criticism of this is that the lack of sunspot numbers was merely a reflection of a lack of observation rather than a true minima. There is another way in which to estimate sunspot activity and that is from auroral activity. Aurorae are caused by high energy charged particles (electrons and protons) emitted by the Sun which are trapped in the upper atmosphere of the Earth. The charged particles are trapped by the Earth's magnetic field in the **Van Allen radiation belts**. The particles are forced along the magnetic field lines towards the Earth's magnetic poles. This is why auroral displays are usually confined to high latitudes. At the magnetic poles the excess electrons and protons discharge. These then collide with atmospheric gas molecules, exciting them and causing them to glow. The greater the solar activity, the more charged particles are emitted and the greater the number of aurorae. Therefore the calmer the Sun, the less aurorae there are. Agnes Clarke had written a note on Maunder's 1894 paper reporting the lack of auroral activity at this time. Eddy points not to the lack of aurorae but the upsurge in reports following the **Maunder Minimum** (Eddy, 1976). Using naked eye observations of oriental sunspots and aurora Eddy identified more than one minimum shown in Figure 4.2. One in the 1300s has been named the Wolf Minimum and one at the start of the Little Ice Age called the Spörer Minimum.

When solar activity is high the magnetic field of the Sun extends. This shields the Earth from some of the cosmic rays, which means less carbon-14 is produced. A quieter Sun increases carbon-14 production due to changes in the solar magnetic field strength of the **solar wind**. On short time-scales of 10 years carbon-14 also varies due to changes in the climate and transfers between the atmosphere and ocean. On long time-scales of 100 000 years the amount of carbon-14 is also determined by the changing geomagnetic pole strength. Stuiver and Quay (1980) looked at the residual carbon-14 variations in tree rings of known date, after the removal of any changes due to

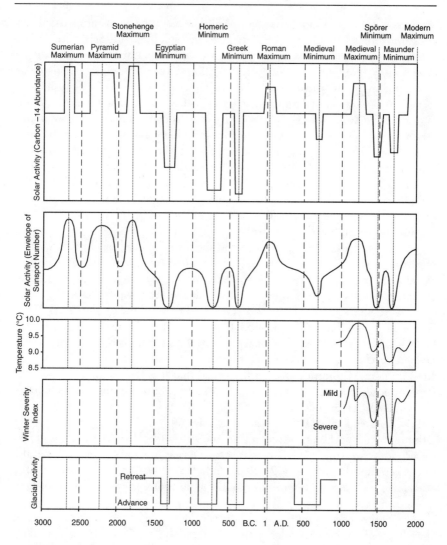

Figure 4.2 Minima and maxima in solar activity
Source: Eddy (1977)

geomagnetic variations. Stuiver and Quay (1980) found major increases in carbon-14 production during 1282–1342 Wolf Minimum, 1450–1534 Spörer Minimum and 1645–1715 Maunder Minimum. These correspond well to the minimum of sunspots and auroral activity. By using carbon-14 activity and the sunspot number Eddy (1977) was able to identify far more solar minima in the historical past (*see* Fig. 4.2). Beryllium-10 is also produced by cosmic rays and has several advantages over carbon-14. First, beryllium-10 only survives in the atmosphere for one to two years compared to 20 years for carbon-14. This allows the solar activity to be well defined. Second, the time resolution for beryllium-10 is high. Third, beryllium-10 is more robust to **anthropogenic** activity compared to carbon-14 (Hoyt and Schatten, 1997). While the beryllium-10 is a good proxy for sunspot numbers for some of the time at low numbers, the time of the Maunder minimum, it reveals solar cycles not shown by the carbon-14 data. This casts some doubt as to its usefulness at low sunspot numbers. It has also been found that the rotational period of the Sun is inversely proportional to sunspots. By using contemporaneous drawings of sunspots and estimating the rotation speed of the Sun, various workers have concluded that prior to the Maunder Minimum the solar rotation rate was slowed and that during the minimum it was substantially faster. Stuiver and Quay (1980) suggest that the increased rotation rate of the Sun inhibits transport of the solar magnetic field to the surface, reducing the solar wind and number of sunspots, but increasing carbon-14 production in the Earth's atmosphere.

The 11-year solar cycle in itself is too short to produce the climatic change required. A longer time-scale process is required to explain the variations in the climate over a century. It is known that there are several other long-term cyclic variations in the Solar activity. The 80-year Wolf-Gleissberg cycle controls the maxima of the 11-year cycle and the lifetime of sunspots. There is a quasi-cyclic fluctuation around 180 years. Other longer cycles have also been suggested, such as a 400-year cycle derived from auroral activity and a 600-year cycle (Hanna, 1996a). This has led some to question the existence of the Maunder Minimum, proposing instead that it is the minimum of several superimposed cycles. The 11-year sunspot cycle is not symmetric. The maximum most often occurs in the first half of the cycle, and rarely in the second half. Gleissberg and Damboldt (1979) found that during the Maunder Minimum the maximum occurred in the second half of the cycle and has never done so since. Rather than the 11-year cycle ceasing to operate, Gleissberg and Damboldt (1979) conclude that it was the asymmetry of the solar cyle together with a minimum in the 80-year cycle that led to lowered sunspot numbers.

The Maunder Minimum does not cover the whole of the Little Ice Age but the observed changes in solar activity correspond well with the global temperature trend. However, there remains the problem of whether a mechanism can be found whereby changes in the solar cycles can alter the Earth's climate. How much does the solar 'constant' vary and is the climate sensitive to these changes? Recent measurements show that the 11-year solar cycle is associated with approximately a 0.1 per cent variability in the Sun's output, the Sun being slightly dimmer at a solar minimum, i.e. when there are less sunspots (Foukal, 1994; NRC, 1994). It is now known that sunspots have associated with them bright areas called faculae and plages which also modify the solar output. They radiate particularly strongly in the ultraviolet and extreme ultraviolet (EUV). Increasing numbers of sunspots means that there are more faculae. The brighter faculae regions are very active and more than compensate for the loss of output from increased sunspot numbers.

A recent solar cycle had maxima in 1980 and 1990 with a minimum in 1986. The solar output dimmed by 0.1 per cent between 1980 and 1986 and then brightened by 0.1 per cent between 1986 and 1990. The estimated climate forcing and consequent surface temperature change resulting from this variation depend on the value of climate sensitivity (equation 2.4). It has been calculated that a 0.1 per cent change in the solar output is equivalent to a climate forcing of 0.24Wm^{-2}. This change seems quite small, but research suggests that the decline in solar output due to the decline in solar activity in the early 1980s was more than enough to offset the predicted net increase in forcing due to anthropogenic activity. However, there are problems with this trade-off. Any surface temperature change is unique to the forcing and depends on the latitude, altitude and history of the forcing. The two forcings cannot be balanced that simply. A further problem is that, roughly, the decline in solar output takes 5.5 years and then it starts to increase again. In that time period the climate system would not have had time to re-equilibrate and so the full expression of the forcing is unlikely to be felt. So while a 0.24Wm^{-2} change in the forcing may lead to an equilibrium change in the surface temperature of 0.1 to $0.2°C$, only a fraction of that would be realized by the transient climate. Important changes on much longer time-scales, those appropriate for the Little Ice Age, have yet to be monitored. It has been suggested that changes in the solar output over 200 years could be as much as 0.25 per cent with a resulting temperature change of $0.46°C$ (NRC, 1994).

There may be ways in which solar variability affects the Earth's climate other than direct effects through changes in solar output. Solar activity affects the solar wind intensity and cosmic ray flux and studies have shown links between these changes and the Earth's climate (Hanna, 1996b). Part of the problem of linking solar cycles to climatic change is that as yet no one can fully explain the variability of solar activity. There is plenty of evidence that our Sun is just like many other stars in the galaxy and varies in output (Hanna, 1996a; Parker 1999). It is now accepted that normal behaviour for a star is to exhibit periods of increased activity as during the Medieval Optimum and periods of decreased activity as during the Maunder Minimum (Parker, 1999). However, finding a mechanism, a direct link, between changes in the solar output and the Earth's climate has proved elusive. Consequently, many of the relationships proposed between the Sun and Earth are based on statistical evidence. These merely provide a correlation with no proven cause and effect (Hanna, 1996b). One of the strongest statistical links between the sunspot cycle and the climate on Earth is the mid-Western drought in the USA which is well correlated with the 22-year cycle (Abbot, 1963). Other links have been made to the El Niño and sea surface temperatures (Mendoza *et al.*, 1991; Reid, 1987). The reliance on statistical relationships between solar cycles and the small changes in solar constant have led to doubts about any significant climate change being caused by solar variations (Houghton, 1994). Others, however, make a strong case for a link between solar variations and climate (Hoyt and Schatten, 1997). Therefore the role of solar variability in climatic change remains a controversial area. The century-scale changes proposed to explain the Little Ice Age are of significance when it comes to considering global warming. Thus interest in this problem remains high, as we shall see later in Chapter 5.

ATMOSPHERIC TURBIDITY

Atmospheric **turbidity** is a measure of the amount of attenuation of solar radiation by the atmosphere. One of the main ways in which atmospheric turbidity can increase is by an injection of dust or aerosols into the atmosphere from a volcanic eruption. Volcanoes are a source of carbon dioxide, sulphur compounds, water vapour and fine dust. Immediately after an eruption the stratosphere is dominated by particles which are 10 times more efficient at scattering light than normal stratospheric particles. Particles can act also as condensation nuclei, increasing cloudiness. In the high troposphere this may initiate cirrus clouds which are persistent and warm the surface. Cooling is thought to dominate, however, as the increase in scattering will cause solar radiation to be reflected into space. The eruptions of Tambora in Indonesia (1815), Krakatoa in Indonesia (1883) and Mt Agung in Bali (1963) have all been associated with a global temperature drop of a few tenths of a degree centigrade. Assessing the effect of any one volcano on the climate is complex and debate still exists over the influence they may have on longer-term climate change.

Like so many of the ideas relating to climate change the notion that volcanoes could affect the climate is not new. Benjamin Franklin observed that volcanic aerosols could reflect sunlight into space and argued that a large eruption off Iceland was responsible for the cold winter of 1783–84. In the 1960s Hubert Lamb devised a measure to calculate the effect the dust from a volcanic eruption would have on the transparency of the Earth's atmosphere (Kelly *et al.*, 1998). Lamb (1970) identified three important attributes of the volcanic eruption:

1. The size of the dust;
2. The height of the dust veil;
3. Lifetime of the dust in the atmosphere.

Using observational evidence Lamb (1970) concluded that the fine dust or ash, 2μm or less across, was the most significant. Observations showed that volcanic dust reached three levels. These were the upper troposphere, the lower stratosphere between 20–27 km and the upper stratosphere with some carried into the mesosphere. In the troposphere the dust is soon washed out by rain. Very little dust reaches the upper stratosphere so its effect is negligible. The dust in the lower stratosphere is the most important. The residence time of aerosols in the stratosphere depends on their size, the height they were injected into the stratosphere and the latitude of the volcanic eruption. The dust survives for years to be carried around the world by the upper winds creating a dust veil. The dust takes 2 to 6 weeks to circle the Earth in middle or lower latitudes and 1 to 4 months to form a complete latitude belt. Volcanoes at low latitudes can potentially cover the whole globe but volcanoes at high latitudes rarely extend lower than $30°$ latitude. This volcanic dust veil acts to stop incoming solar radiation, because as the particles are typically about the same size as solar wavelengths they reflect the radiation into space. Using observational evidence of the effect of dust veils on the direct radiation beam, Lamb designed the **dust veil index** or DVI. The eruption of Krakatoa in 1883 was designated 1000 (Kelly *et al.*, 1998). It might be thought that the more explosive the volcano the greater the effect the eruption would have on the climate. This is not always the case, however, as the eruption of St Helens in the USA in 1980 showed. This very powerful eruption produced only a small DVI.

Temperatures below freezing leave a mark in tree rings which also indicates whether the freezing occurred at the beginning or end of the growing season. Frost-rings can be found in trees within two years of the eruptions of Krakatoa in 1883, Mt. Pelée and La Soufiere (West Indies) in 1902, Katmi (Alaska) in 1912 and Mt. Agung in 1963 (Grove, 1988). These provide evidence of the link between climate and volcanic eruptions. Data from Bristlecombe pines have been correlated with Lamb's DVI. The correlation between frost events and volcanic events was found to be six times that expected by chance (Grove, 1988). In the data record, however, frost rings occur without volcanic events and vice versa. It is now thought that it is sulphur-rich volcanic eruptions that cause climate change. These release vast quantities of sulphur dioxide into the atmosphere which form small droplets of sulphuric acid. These aerosols have a far longer residence time in the atmosphere than the dust. *In situ* stratospheric measurements show that the dominant aerosol is sulphuric acid. Thus the potential climatic impact of a volcano is not only governed by its magnitude but also by the chemical composition of the magma, that is the concentration and type of volatile that may be outgassed during an eruption (Carroll, 1997). The lower the sulphur content, the less an effect the eruption will have on the climate and this may explain why some eruptions produce no frost ring. A measure of the sulphur content of the atmosphere comes from the acidity of annual layers in ice cores extracted from polar ice sheets. The oxygen 18 isotope provides a way of dating the core. A core from Greenland which has been dated to within a year, reveals many eruptions that were known about, but also many that were not. Satellite observations indicate that between one third and one quarter of volcanic eruptions occur in such isolated parts of the world that they would not be recorded.

The Mt Pinatubo eruption in the Philippines in June 1991 provided a great deal of observational evidence to back current theory. An individual volcanic eruption would appear to cause a temperature decrease of a few tenths of a degree within several months of a major event. However, the effect is very dependent on the hemisphere. The northern hemisphere reacts quickly to eruptions that occur in that hemisphere, but more slowly to ones in the southern hemisphere. The southern hemisphere shows little response to northern hemisphere eruptions and reacts slowly to ones in the south. This is because the southern hemisphere is dominated by oceans which damp down any change (Kelly *et al.*, 1998). The effects of one eruption may upset the climate for a year or so. A clustering of several volcanoes would be necessary to have any long term effects on the climate. Groups of volcanic eruptions have been put forward as an explanation for the Little Ice Age. Ice cores reveal a good correlation between periods of below-average volcanic activity and above-average temperature. The medieval warm period coincides with a time of little volcanic activity, from 1100 to 1250. In contrast, some of the coolest periods of the Little Ice Age, between 1250 to 1500 and 1550 and 1700, occur when there was a lot of volcanic activity. There were more eruptions in the early twentieth century and 1970s than between 1925 and 1955. This coincides with rising global surface air temperatures between the 1920s and 1940s followed by the glacial advance of the 1960s and 1970s. However, the group of eruptions between 1881 and 1889 had no greater effect than one large eruption (Grove, 1988). Thus the effect of volcanoes on decadal time-scales remains difficult to determine.

Internal forcings

It has been proposed that natural variability of the climate system may explain many of the observed short-term climate changes that have previously required an external climate forcing to be grossly amplified. Stocker and Mysak (1992) put forward the argument that climatic variations on century time-scales can be attributed to changes in the thermohaline circulation of the Pacific–Atlantic basin. This could explain the Little Ice Age without recourse to changes in solar output. The Little Ice Age could be due to small fluctuations in the deep water circulation over periods of 50 to 200 years, causing surface temperature variations of 0.1 to 0.5°C. This theory is very similar to that proposed by Broecker and Denton (1990) to explain the glacial cycles, except that it is on much shorter time-scales. Many of the changes which occur within the climate system are on shorter time-scales still. The El Niño, discussed in Chapter 2, is the classic example of the interaction between sea surface temperatures and atmospheric circulation which occurs on time-scales of 2 to 5 years. Sea surface temperature anomalies are known to accompany changes in the atmospheric circulation. Such changes may persist for several years. Indeed, it seems well established that warm sea surface temperatures in the Atlantic lead to exceptionally warm summers and vice versa.

The ice–albedo feedback can influence the climate on short time-scales. Changes in Arctic sea-ice affect the atmospheric circulation and vice versa. This is particularly noticeable over Europe. A change from ice to open water has a strong effect on the heating of the overlying atmosphere. Any strong thermal gradients in the atmosphere shift with the sea-ice extent. Jet streams and depression tracks tend to be displaced parallel to the ice limit and this is particularly strong in the North Atlantic. The slackness or lack of vigour in the atmospheric circulation can also be responsible for sea-ice growth, just as vigorous circulation causes a retreat. This occurred in the 1920s and 1930s in response to atmospheric circulation (Lamb, 1982; Grove, 1988).

The atmosphere can reflect the thermal conditions of the underlying land surface when large areas are either frozen or flooded. For example, in Antarctica the great frozen plateau is extremely cold and the overlying air likewise. Incursions of warm air do little to change the temperature. Whether changes in the land surface can affect the general circulation is debatable, however short-lived anomalies do occur. For Britain the summer of 1976 was extremely hot and one of drought conditions. The summer of 1976 is noted for the persistence of an anti-cyclone over Britain. Blocking highs are well known but this one was very long-lived. It has been suggested that the drought conditions and parched land reinforced the anticyclone, further maintaining the hot conditions.

The importance of natural climate change

Some of the debate surrounding global warming focuses on whether any of the forcings described in this section are large enough or occur on sufficiently long time-scales to influence the long-term impact of increased greenhouse gases. Nevertheless these forcings are important to understand, for any changes they cause in the climate will be superimposed on the global warming signal. In some cases they may even mask the effect of increasing greenhouse gases, at least for a limited period. They are also important because it is predominantly these natural variations that humankind have had to respond to in the recent past. How have these small variations affected societies in the past, and can they offer us any insights into the future?

Climate and human history

As Ingram *et al.* (1981) report, there is widespread disagreement on whether the climate can truly be said to have influenced human history and society. There are several reasons why historians have not been interested in climate change as an area of study. Nowadays climate change is taken as a matter of fact, but that was not always the case. The concept of a changing climate has gone in and out of fashion and if climate does not change it cannot possibly affect human history. Indeed, early observational evidence suggested that climate did not change. Climate change was thought to be too small to cause social and economic change. It was thought that the time-scale for any significant climate change was too long to affect humankind. There was a minority of researchers, principally archaeologists and natural scientists with a few historians, who believed otherwise. The extremists in this group were the 'climatic determinists'. For these researchers climate was one of the most important factors to influence history and the development of civilization. Huntington (1907) and Brooks (1949) were amongst the most influential in this field. While these views were very popular in the first half of the twentieth century, they fell from favour in the 1950s (Pearson, 1978). Partly this was due to a perceived lack of scientific rigour but also the rather extreme stance taken by some researchers. In the 1970s the ideas were revived by Lamb (1982) and, with a more credible scientific approach emerging, the idea of a link between climate change and human history became acceptable again (Grove, 1988). However, it is only in the past decade that these ideas have once more commanded centre stage. Take for instance the diagram of the climate system as shown in the 1990 Intergovernmental Panel on Climate Change (IPCC) reproduced in Figure 1.12. Now compare it to the diagram in the 1995 IPCC report shown in Figure 4.3. This reflects our new found concern at the effect of human activities on the Earth's environment. Whether naturally occurring climate change can influence human history still remains a source of debate. The major difficulty in assessing the extent of the effect of climate on human society is the complexity of climate–history inter-actions and the lack of detailed information on past weather and climate.

Part of the problem extends to proving a link. Very often studies have assumed a link just because two events are coincident. The example given by Ingram *et al.* (1981) for this is the space–time coincidence. Because the Roman Empire fell at a time of cold weather does not mean climate caused the downfall of Rome. Cause and effect have not been proven. Even the multiplicity of events that can be pointed out as coinciding with climatic extremes can be countered with contrary examples (Lamb, 1982). There is also the problem of data. Trying to link climate and history requires considerable observational data. Prior to the 1700s this was not available, and proxy climatic data have to be used. Therefore much of the data that have to be used are influenced by other variables connected to economic, political and social developments as well as climate. Grain production has been widely used as a climatic indicator, but even this only relates to a specific area at a time when measures were not uniform (Parry, 1978). Much of the data are only available for Europe, which may be unrepresentative. Simple correlations are not enough to prove cause and effect. Equally, just because there is no correlation does not prove there is no climatic influence. De Vries (1980) found that there was no correlation between winter temperatures and butter prices in Leiden Holland between 1635 and 1839. However, no account was made of other meteorological factors and the periods of the year significant to dairy farming. There was a statistically significant correlation between late frosts in March and butter price rises.

Ideally therefore in order to prove a link between climate and human history

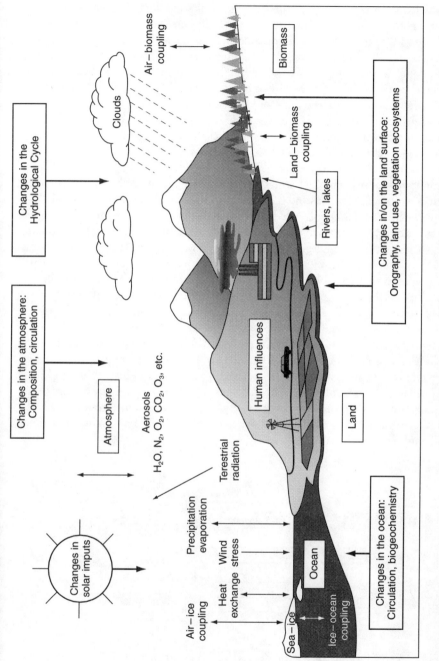

Figure 4.3 The latest IPCC representation of the global climate system, note the incorporation of human activity
Source: Houghton *et al.* (1996)

one needs to develop a model. This model would consider all the processes involved in possible climate-history links. Figure 4.4 shows one of the simplest which only points out the possible climatic links and does not include any other factors that may affect human activity. Ideally one should build an econometric model taking into account economic, social and climatic variables and specifying the links between them. However, the data are unlikely to be available and so complex are the interactions that Ingram *et al.* (1981) suggest that a small model is more likely to be successful. A model which concentrates on a few variables, on the first- and second- order climatic links and on a specific geographical area. Other strategies to link history and climate include studying certain space–time events in particular looking at a society during a period of climate stress which is associated with marked changes in the economic and political structure. This combines a modelling approach with a qualitative approach. The study looks in detail at the relationship between meteorological variables and economic variables and then, when the modelling approach can no longer be used due to lack of data, turns to a more descriptive approach. However, such studies may well over-emphasize the role of climate in society. Another method suggested by Ingram *et al.* (1981) is Occam's Razor. Occam was an early fourteenth-century English philosopher who put forward the maxim that in explaining something assumptions must not be needlessly multiplied. For example, how can we explain the appearance of crop circles? We could assert they are made by extraterrestrials but then we have to assume they exist, we assume they have spaceships to visit us and that they want to communicate with us in a very bizarre way. Alternatively we could assume crop circles are made by someone with a sense of humour. Assumption should not be piled upon assumption just to fit together climate and history. If a political upheaval has taken place and it can be explained most simply by factors other than climate, then those are more likely explanations. This method is highly subjective but can be used to remove instances where climate change was of little importance (Ingram *et al.*, 1981).

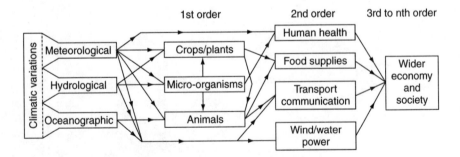

Figure 4.4 Difficulties of linking climatic variations to increasingly more removed impacts
Source: Ingram *et al.* (1981)

Case Studies

There are various events, usually in military history, that illustrate how short-term climatic events, really meteorological phenomena, have influenced the course of history. In the late eighteenth century France hoped to attack England by inciting rebellion in Ireland. The French fleet was to land at Bantry Bay in December 1796. While initially fog protected the French fleet from the English it also prevented efficient communication amongst the fleet. There then followed a period of unsettled weather which prevented the French from landing in Ireland and effectively stopped any further thoughts of invasion (Wheeler, 1998). It is difficult to assess the real long-term effect that these individual events have. Wheeler (1998) points out that the French fleet soon recovered to pose a threat to England. A more important climatic influence may be seasonal effects on agriculture. Such variations may help explain the demise of the first English colony in America in 1587 and the difficulties of the second in 1607. The failure of the colonies at Ranoke Island, North Carolina and Jamestown, Virginia has traditionally been blamed on poor management and hostile natives. Recent work suggests that climate may play a role as both colonies were founded at times of severe drought. In both cases it was the worse drought for nearly 800 years (Stahle *et al.*, 1998). The local population was reluctant to trade food with the colonists because drought had depleted their crops. This exacerbated the problems for the settlers and increased friction between them and the locals.

One of the most well-documented studies of the potential longer-term influence of climate change on human history is that of the Norse Greenlanders (Barlow *et al.*, 1997; McGovern, 1981; Pringle 1997). It is the study of two settlements in Greenland founded during the medieval optimum and their decline during the Little Ice Age. The Norse colonies in West Greenland existed from 985 to 1500 AD. It was the westernmost outpost of Atlantic Europe at that time (*see* Fig. 4.5). The Norse people had prolonged contact with the Thule hunters; the forebearers of the modern day Inuit. The Inuits were also settlers, having arrived in 1100 AD from Ellesmere Island (Pringle, 1997). The Inuits became efficient hunters and continued to thrive. In contrast, the Norse farmer colony perished. In order to explain the disappearance of the Norse settlements a complex interweaving of social, economic and climatic evidence has been used by researchers (Barlow *et al.*, 1997; McGovern, 1981; Pringle 1997). The evidence of the economic and social history of the Norse settlements comes from various fragmented documentary sources. Work also comes from archaeological evidence. The climatic evidence comes from studies of the Greenland Ice Sheet Projects, GISP 1 and 2, and population cycles in the flora and fauna exploited by the Norse settlers.

The west coast of Greenland has two climatic regimes; the oceanic coastal zone is very wet but at the heads of the deep fjords a continental climate exists. The latter is dry but florally lusher than the coastal zone. In 982 AD according to the Greenlander Saga, Erik the Red discovered the west of

Greenland, naming it so to encourage others to follow him, not because it
was Green. However, as pointed out by Dansgaard *et al.* (1975), it may have
been green at the time because it was warmer (as shown by oxygen isotope
measurements). Other documentary evidence supports the notion that the
average water temperature of the fjords was 4°C warmer than present (Lamb,
1982). The Norse colonists arrived from Iceland and created two settlements.
The first, which was more northerly but on the west coast, is called the western settlement or Vesterbygd. The second settlement was founded in 985 AD
and is called the eastern settlement or Østerbygd (Figure 4.5). They settled
inland in the inner area with its continental climate. The east settlement was
estimated to have some 4000–5000 inhabitants and the more northerly western settlement had 1000–1500 settlers. The western settlement farms were
associated with the lowland pastures of the inner fjords and the lush biota.
The pasturing of cattle, sheep and goats determined the farm's position as
grain agriculture was not successful in Norse Greenland. It appears from
archaeological evidence that the Norse also relied on hunting migratory Harp
seals and Caribou. Even those farms several hours from the fjordside were

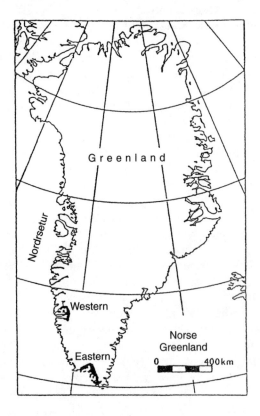

Figure 4.5 The western and eastern settlement of the Norse
Vikings in Greenland
Source: McGovern (1980)

highly dependent on seals. Surprisingly, given their proximity to the sea, the evidence suggests that they did not fish (Pringle, 1997). McGovern (1981) describes a seasonal cycle whereby during the long Greenlandic winter the Norse relied on cattle shut in heavily insulated byres. The cattle would have to stand in their own dung and live off stored fodder. By mid-winter milk production dropped leaving the Norse community reliant on stored meat and diary produce. The arrival of the Harp seals to the outer fjord areas in spring, during May and June, would have been of vital importance even during good climatic conditions (*see* Fig. 4.6).

As well as the seasonal hunt and hay gathering there were long hunting trips to Disko Bay 800 km north of the west settlement. These occurred each year in summer and the Norse hunted for polar bears and walruses. The hunts brought back tusks and hides vital for trade with Europe. Contact with Europe was irregular though and the voyage from Europe was long. This made famine relief from Europe out of the question. Also the boats were small and could not have carried much food (Grove, 1988). Therefore the Norse Greenlanders did not depend on this European trade. However, McGovern (1981) reported that European goods were obviously highly prized by the Greenlanders. Their economy was precarious, relying on communal labour and seasonal practices. They did not not use the inner-fjord resources to full potential. The tight schedule was best suited to highly predictable seasonal and year-to-year resource fluctuation. This pattern operated well for 150 years (*see* Fig. 4.6). The Norse Greenland political and social structure was derived from Iceland with the first settlers getting the best land (*Landnamsmen*) and the followers (*thingmen*) getting what they were given. There was a chieftain–retainer relationship and a hierarchical organization developed. The Church also grew in power and built some spectacular churches. The trade with Europe provided the stained glass, bell iron and even a bishop. Greenland probably became as highly structured as Iceland with power concentrated in a lay and clerical elite with a two- or three-tiered management structure.

The deteriorating weather conditions that took hold as the Little Ice Age approached would have significantly affected the dual maritime–terrestrial economy. The terrestrial component of the Norse economy would have been most affected by increasing precipitation in the inner fjords. In the winter snow would have kept the cattle in the byres for longer and made the unsanitary conditions worse. In summer the amount of hay stock for winter would have been reduced. During the irregular cooling of 1250–1300 and the 1300s, Harp seal migrations were less regular and therefore less accessible to the hunters from the colony. Oceanic storms may have penetrated the inner fjords causing further problems for the terrestrial economy. As the climatic conditions deteriorated, the Thule Innuit migrated southward along the west coast. This brought the Innuit and Norse huntsmen into increasing competition and it seems likely that the two sides fought over the prime sealing areas. The links with Europe also became less and official contact ended after 1408. Increasing

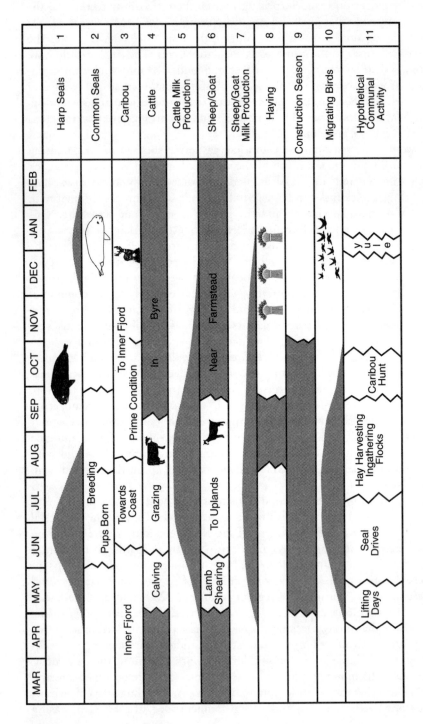

Figure 4.6 The yearly activities of the Norse communities
Source: McGovern (1980)

drift ice in the Denmark Strait and Danish–Norwegian economic and political realignments both played their part in the diminishing trade (Grove, 1988; McGovern, 1981).

The Norse colony disappeared due to changes in the climate, politics and economics, had they not all coincided the Norse colony may have survived. But this solution ignores the possibility of human response to climate change. The success of the Innuit shows that the deteriorating climate did not make Greenland inhabitable. The Norwegian settlers in Finnmark (the furthest province in modern Norway) thrived in the deteriorating climatic conditions (Brown, 1994). Had the Norse settlers chosen to abandon their inner fjord pastures for the coastal oceanic zone and a more maritime economy they might have survived. However, archaeological evidence shows they persisted with their settlement pattern and their seasonal dual economy. They built larger byres and store houses to sustain them through the winter. The Norse did not take on other skills such as skin boat building (rather than the traditional wooden boats) which may have helped them exploit other resources such as the Ringed seal colonies. The Norse stuck to European fashions, rather than the skin clothes of the Innuit. In contrast, the Innuit used what part of Norse culture they required. The Norse society continued 'elaborating its churches rather than its hunting skills' (McGovern, 1981). In 1342 when Bardsson took up stewardship of the West colony there were 90 farms and 4 churches but by about 1350 it was deserted. GISP2 ice core data show that some of the coldest years were between 1343 and 1362. Some researchers believe the collapse was catastrophic. The west colony began eating their cows, which was forbidden under Norse law, and their hunting dogs before finally starving to death. Others suggest a less harrowing end with the western settlers packing their bags and moving to the eastern settlement (Pringle, 1997). The East colony disappeared more slowly but was gone by 1500. Whaling by Europeans had become more popular and it is possible that an already weakened East colony was attacked by pirates. The pirates may also have carried the plague which is known to have arrived in Iceland in 1402 (Grove, 1988). Evidence for a gradual deterioration in health comes from the average height, which is often taken as an indicator of nutritional intake. The average height of the men buried in the settlement declined from 5' 10" when they first arrived to 5' 5" by its end (Lamb, 1982).

When the Norse first arrived they had shown tremendous flexibility and ingenuity, modifying the Icelandic economic model to their own needs. Later on, however, they appeared unable to make the neccessary changes required to cope with the deteriorating climate and economic stresses. McGovern (1981) blames this increasing inflexibility on the growth of hierarchy and elitism within the Norse community. When there is little stress on a society there is no need to gather data, or to manage. As a society comes under climatic or economic stress, there is a need for more data. However, there is a limit to how much data one person can process. People begin to specialize and the hierarchy becomes more stratified. The more stress, the more elaborate becomes the hierarchical structure to process the environmental and economic data. This increases the distance between social classes. Decisions will be made which benefit the elite and not necessarily the individual farmer. In order to justify this an increasing mystification of the elite was needed and this was probably achieved by increasing the power of the church. As the distance between farmer and elite grows, there is a danger that management effort goes towards maintaining the hierarchy, increasing conformity and ritual. If the elite managers perform well, searching out additional resources and alternative technologies, such as that of the Innuit, this separation can work to the benefit of the society as a whole, but the elite in Greenland clung to their rituals. McGovern (1981) suggests that a society which believes that lighting more candles to St Nicholas will have a greater effect on the spring seal hunt than more and better boats is in deep trouble. The last records from Norse Greenland are of a Christian marriage and a burning for witchcraft.

What this study shows is that climatic, economic and political change can all put stresses on a society which may threaten its survival. What is equally important, however, is the response to these stresses. It might be argued that western Europeans have a superior attitude to other cultures. The Norse settlers spurned Thule culture preferring to cling to Norwegian values, even if it meant dying in the process (Pringle, 1997). There are studies that reveal other cultures which also appeared more concerned with maintaining traditions than adapting to a changing environment. The Mayan civilization on the Yucatan Peninsula in central America flourished between 300 to 800 AD (Lamb, 1982). Despite the dense tropical rainforest they were an agriculture-based society. Using a sophisticated network of drainage and irrigation channels they farmed extensively in the lowlands of Guatemala and Honduras. The forest at this time was easier to keep at bay as the area was drier than it is now. Climatic reconstructions from oxygen isotope and sediment core analysis from Lake Chichancanab, Mexico show that between 800 and 1000 AD an even drier climatic regime set in (Hodell *et al.*, 1995). By 1000 the Mayan society collapsed. It would appear that the society was tied in to a particular agricultural system that could not respond when the climate changed (Pain, 1994). Climate was only part of the problem. The society's inability to change, to acknowledge the real problem was the reason they did not survive. They continued to consume on the same level they always had. Sounds familiar? In our own global society the response to stresses may be equally important when we come to tackle global warming and the potential economic and political change.

Summary

Climate change in historical times challenges some of the modern preconceptions of our supremacy over the natural world. It questions whether our civilizations and societies are as stable as we like to think. While our society becomes more and more able to cope with short-term extreme weather events, society's ability to cope with major climatic events may decline due to its very elaborate structure (Wigley *et al.*, 1981). Perhaps, then, it is not surprising that the influence of climate change on society remains a contentious issue and sadly, until recently, a slightly neglected one. It falls between so many disciplines and is often the prerogative of a lone researcher rather than a large group. Many facets of potential studies into the issue are unattractive to those that would fund scientific research, yet it can provide fascinating insights as to how humans have struggled not just with climate change but the mores of their society. If global warming is to come then these are the challenges that we face, not just the temperature and precipitation changes, but also our perceptions of them and our reactions to them as a society.

Chapter 2 noted how the early work on the greenhouse theory in the late eighteenth century had been carried out to find a reason for the quaternary glacial cycles. The greenhouse theory, however, was abandoned by the early twentieth century. It was thought that water vapour absorbed most of the outgoing infra-red radiation in the region of the carbon dioxide absorption bands. Therefore the effect of changing levels of carbon dioxide was immaterial. Also, there was the perception that climate, at least on historical time-scales, was constant. It was a scientist called Callender who revived the debate in 1938. He made a direct link between rising carbon dioxide levels from anthropogenic activity and increasing global temperatures (Callender, 1938). This view was dismissed and heavily criticized. Although world-wide meteorological measurements now pointed to a world with a changing climate, the concept that humans could change nature in this way seemed remarkable. Furthermore, it was widely accepted that any excess carbon dioxide produced by industrial activity would be absorbed by the oceans. In 1957 this view was challenged in an article by Revelle and Suess. They concluded that the oceans were not absorbing all the excess carbon dioxide and, although the biosphere was absorbing some carbon dioxide, levels in the atmosphere would rise by 20 to 40 per cent in the next few decades. This led to the establishment, in 1958, of the regular monitoring of atmospheric carbon dioxide levels at Mauna Loa in Hawaii. Since then, although global warming has remained a hotly contested issue, its importance as a scientific and political concern has continued to grow in leaps and bounds.

References

Abbot, C.G. 1963: *Solar variation and weather.* Smithsonian Miscellaneous Collections 146, no. 99. Washington, DC: Smithsonian Institute.

Barlow, L.K., Sadler, J.P., Ogilvie, A.E.J, Buckland, P.C., Amorosi, T. Ingimundarson, J.H., Skidmore, P,. Dugmore. A.J. and McGovern, T.H. 1997: Interdisciplinary investigations of the end of the Norse western settlement in Greenland. *Holocene* 7, 489–99.

Bradley, R.S. and Jones, P.D. (eds) 1992: *Climate since AD 1500.* London: Routledge.

Brazell, J.H. 1968: *London weather.* London: Meteorological Office.

Broecker, W. and Denton, G.H. 1990: What drives glacial cycles. *Scientific American* 262, 43–50.

Brooks, C.E.P. 1949: *Climate through the ages.* 2nd edition. London: Ernest Benn.

Brown, N. 1994: Climate change and human history. Some indications from Europe 400–1400 AD. *Environmental Pollution* 83, 37–43.

Callendar, G.S. 1938: The artificial production of carbon dioxide and its influence on temperature. *Quarterly Journal of the Royal Meteorological Society* 64, 223–37.

Carroll, M.R. 1997: Volcanic sulphur in the balance. *Nature* 389, 543–4.

Dansgaard, W., Johnsen, S.J., Reeh, N., Gundestrup, N., Clausen, H.B. and Hammer, C.U. 1975: Climatic changes: Norsemen and modern man. *Nature* 255, 24–8.

Duncan, K. 1991: A comparison of the temperature records of Edinburgh and central England. *Weather* 46, 169–73.

Eddy, J.A. 1976: The Maunder Minimum. *Science* 192, 1189–202.

Eddy, J.A. 1977: The case of the missing sunspots. *Scientific American* 236, 80–92.

Foukal, P. 1994: Study of solar irradiance variations holds key to climate questions. *Eos* 75, 377.

Gleissberg, W. and Damboldt, T. 1979: Reflections on the Maunder Minimum of sunspots. *Journal of British Astronomical Association* 89, 440–9.

Grove, J.M. 1988: *The Little Ice Age.* London: Methuen.

Hanna, E. 1996a: Have long-term solar minima, such as the Maunder Minimum, any recognisable climatic effect? Part 1 Evidence for solar variability. *Weather* 51, 234–42.

Hanna, E. 1996b: Have long-term solar minima, such as the Maunder Minimum, any recognisable climatic effect? Part 2 evidence for a solar-climatic link. *Weather* 51, 304–12.

Hecht, A.D. (ed.) 1985: *Paleoclimate analysis and modeling.* New York: John Wiley and Sons.

Hodell, D.A., Curtis, J.H. and Brenner, M. 1995: Possible role of climate in collapse of Classic Maya civilization. *Nature* 375, 391–4.

Houghton, J. 1994: *Global warming: the complete briefing.* Oxford: Lion Publishing.

Houghton, J.T., Meira Filho, L.G., Callander, B.A., Harris, N., Kattenberg, A. and Maskell, K. (eds) 1996: *Climate change 1995: the science of climate change.* Cambridge: Cambridge University Press.

Hoyt, D.V. and Schatten, K.H. 1997: *The role of the Sun in climate change.* Oxford: Oxford University Press.

Hulme, M. and Jones, P.D. 1994: Global climate change in the instrumental period. *Environmental Pollution* 83, 23–36.

Huntington, E. 1907: *The pulse of Asia.* Boston: Houghton Mifflen.

Ingram, M.J., Farmer, G. and Wigley, T.M.L. 1981: Past climates and their impact on man: a review. In Wigley, T.M.L., Ingram, M.J. and Farmer, G. (eds), *Climate and history: studies in past climates and their impact on man.* Cambridge: Cambridge University Press.

Jones, P.D. and Bradley, R.S. 1992: Climatic variations over the last 500 years. In Bradley, R.S. and Jones, P.D. (eds), *Climate since AD 1500.* London: Routledge.

Kelly, P.M., Jones, P.D., Robock, A. and Briffa, K.R. 1998: The contribution of Hubert H. Lamb to the study of volcanic effects on climate. *Weather* 53, 209–22.

Kington, J. 1999: The severe winter of 1697/98. *Weather* 54, 43–9.

Lamb, H.H. 1970: Volcanic dust in the atmosphere; with a chronology and assessment of its meteorological significance. *Philosophical Transactions of the Royal Society, Series A Mathematical and Physical Sciences* 266, 425–533.

Lamb, H.H. 1982: *Climate history and the modern world.* London: Methuen.

Manley, G. 1974: Central England temperatures: Monthly means 1659–1973. *Quarterly Journal of the Royal Meteorological Society* 100, 389–405.

McGovern, T.H. 1980: Cows, harp seals and churchbells: adaptation and extinction in Norse Greenland. *Human Ecology* 8, 245–75.

McGovern, T.H. 1981: Economics of extinction in Norse Greenland. In Wigley, T.M.L., Ingram, M.J. and Farmer, G. (eds), *Climate and history: studies in past climates and their impact on man.* Cambridge: Cambridge University Press.

Mendoza, B., Perez-Enriquez, R. and Alvarez-Madrigal, M. 1991: Analysis of solar-activity conditions during periods of El Niño events. *Annales Geophysicae* 9, 50–4.

NRC 1994: *Solar influences on global change.* Board on global change commission on geosciences, environment and resources, National Research Council. Washington DC: National Academy Press.

Pain, S. 1994: 'Rigid' cultures caught out by climate change. *New Scientist* 141 (no. 1915), 13.

Parker, E.N. 1999: Solar physics – sunny side of global warming. *Nature* 399, 416–7.

Parry, M.L. 1978: *Climatic change, agriculture and settlement: studies in historical geography.* Folkestone, CT: Archon Books.

Pearce, F. 1996: Lure of the rings. *New Scientist* 152 (no. 2060), 38–40.

Pearson, R. 1978: *Climate and evolution.* London: Academic Press.

Pfister, C. 1992: Monthly temperature and precipitation in central Europe from 1525 to 1979: quantifying documentary evidence on weather and its effects. In Bradley, R.S. and Jones, P.D. (eds), *Climate since A.D. 1500.* London: Routledge.

Pringle, H. 1997: Death in Norse Greenland. *Science* 275, 924–6.

Revelle, R. and Suess, H.E. 1957: Carbon dioxide exchange between atmosphere and ocean and the question of an increase of atmospheric CO_2 during the past decades. *Tellus* 9, 18–27.

Reid, G.C. 1987: Influence of solar variability on global sea surface temperatures. *Nature* 329, 142–3.

Spellman, G. 1999: Wine, weather and climate. *Weather* 54, 230–9.

Stahle, D.W., Cleaveland, M.K., Blanton, D.B., Therrell, M.D. and Gay, D.A. 1998: The lost colony and the Jamestown droughts. *Science* 280, 564–7.

Stocker, T.F. and Mysak, L.A. 1992: Climatic fluctuations on the century time scale: a review of high-resolution proxy data and possible mechanisms. *Climatic Change* 20, 227–50.

Stuiver, M. and Quay, P.D. 1980: Changes in atmospheric carbon-14 attributed to a variable sun. *Science* 207, 11–19.

Thompson, L.G., Mosley-Thompson, E., Dansgaard, W. and Grootes, P.M. 1986: The Little Ice Age as recorded in the stratigraphy of the tropical Quelccaya ice cap. *Science* 234, 361–4.

Vries, J. de 1980: Measuring the impact of climate on history: the search for appropriate methodologies. *Journal of Interdisciplinary History* 10, 599–630.

Wheatley, H.B. (ed.) 1906: *Diary and correspondence of John Evelyn with life 1648–85*. London: Bickers and Son.

Wheeler, D.A. 1998: The Bantry Bay incident, December 1796: an example of climatic influences in the age of sail. *Weather* 53, 201–9.

Wigley, T.M.L., Ingram, M.J. and Farmer, G. (eds) 1981: *Climate and history: studies in past climates and their impact on man*. Cambridge: Cambridge University Press.

|5|

The observational evidence for global warming

Introduction

By the 1980s concern about global warming had grown to such a level that the world's governments decided to take at least some monitoring action. In 1988 the Intergovernmental Panel on Climate Change (IPCC) was set up to assess the scientific information regarding global warming and also to suggest response strategies. Its first report considering the science of global warming became available in 1990 (Houghton *et al.*, 1990) and the latest was in 1995 (Houghton *et al.*, 1996). A new report is due at the time of writing this book (2000). These reports are the considered views of hundreds of scientists around the globe, they have undergone considerable criticism and have been rewritten many times. The next two chapters will therefore provide only a short summary of the wealth of detail to be found in these reports. The purpose of these chapters is provide a background to the reports and in doing so explain the main difficulties in the science and the problems of interpreting the results. These two chapters will also provide some of the criticisms of the science which a *minority* of scientists feel calls into question the whole notion of global warming and which they also feel has not been adequately addressed by the reports. It must be stressed that this is a minority view but it has been influential in casting doubt on global warming and in affecting action to prevent global warming. Consideration of the wider consequences of this vocal minority will occur in Chapter 8. For now we will stick to the science.

Much of the evidence that exists to support the claim that global warming will occur, particularly what sort of changes will happen to the climate, is in the form of model data. The models and model data will be considered in the next chapter. This chapter looks at some of the commonly asked questions of the observed data that relate to global warming, the most obvious one being, does the observational data show that global warming is happening? The chapter will also explore some of the criticisms of the observed data and begin to reveal the debate that has set global warming alight.

The greenhouse gases

It may have taken a trip round the solar system in Chapter 3 but by now you should be convinced that carbon dioxide is a **greenhouse gas** and is of considerable importance. However, so far we have really only considered the natural **greenhouse effect** which, as Chapter 3 showed, for planet Earth has been very beneficial, allowing life to evolve and keeping it at just the right temperature. Now we need to look at which greenhouse gases are being increased above pre-industrial levels. We also need to know which live long enough in the atmosphere to potentially cause global warming and how efficient the gas is at absorbing long-wave radiation.

Table 5.1 Levels of the main greenhouse gases compared to pre-industrial levels

	Carbon dioxide (C0$_2$)	Methane (CH$_4$)	Nitrous oxide (N$_2$O)	CFC-11
Pre-industrial concentration	~280 ppmv	~700 ppbv	~275 ppbv	zero
Concentration in 1994	358 ppmv	1720 ppbv	312 ppbv	268 pptv
Percentage increase per year	0.4	0.6	0.25	0
Atmosphere lifetime (years)	50–200	12	120	50

Source: Houghton *et al.* (1996)

The 1990 IPCC report (Houghton *et al.*, 1990) and subsequent reports identified the main greenhouse gases that have increased since pre-industrial times and that would contribute to global warming. The major greenhouse gases are carbon dioxide, followed by chlorofluorocarbons (CFCs), methane (CH$_4$) and Nitrous Oxide (N$_2$O). Table 5.1 shows how these greenhouse gases have been rising from what are considered to be pre-industrial levels. CFCs are part of a larger group of compounds called halocarbons. These substances contain carbon and members of the halogen family which is composed of fluorine, chlorine, bromine and iodine. Halocarbons that include chlorine and bromine are greenhouse gases. They are also capable of depleting the stratospheric ozone layer. Stratospheric ozone depletion is often identified as the primary cause of global warming, but this is not the case. The cause of global warming is increasing greenhouse gases in the atmosphere. The relationship between stratospheric ozone depletion and global warming is complex and will be dealt with later in this chapter. Tropospheric ozone (O$_3$) has also increased and, as ozone is a greenhouse gas, can lead to warming. However, the case of ozone is much more complex than the previous four gases mentioned.

The amount of a gas in the atmosphere depends on the balance between its sources and sinks. Sources arise from emissions of the gas directly into the atmosphere and also production of the gas in the atmosphere. Sinks occur due to destruction of the gas in the atmosphere and uptake of the gas by the Earth's surface. These processes operate on different time-scales but once combined together give a lifetime of a gas in the atmosphere. The following summarizes the natural sources and sinks of the most important greenhouse gases and identifies the important anthropogenic sources with current estimates for concentration increases (*see* also Fig. 5.1).

Carbon dioxide

In the moist lower atmosphere water vapour is the most efficient absorber of long-wave radiation, but the concentration of water vapour will not be

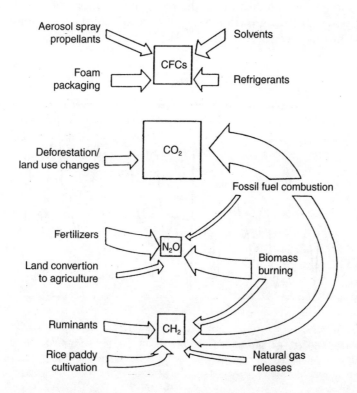

Figure 5.1 The anthropogenic sources of the four main greenhouse gases. The size of the boxes represent their relative contribution to global warming and the arrow size indicates the importance of that anthropogenic source
Source: Warrick *et al.* (1990)

directly affected by anthropogenic emissions (Houghton *et al.*, 1990). Carbon dioxide is the next most important absorber of long-wave radiation. Observations of the amount of carbon dioxide in the atmosphere are monitored at various stations around the world. The most famous of these is at Mauna Loa on Hawaii where records extend back to 1958. Figure 5.2 shows the growth rate of carbon dioxide in parts per million volume (ppmv) per year and reveals a general upward trend. The sawtooth superimposed on this upward trend is the seasonal variation in carbon dioxide. The amount of carbon dioxide in the atmosphere decreases during northern hemisphere summer. It is thought that this is due to the increased northern hemisphere land biomass taking up more carbon dioxide at this time (Hall *et al.*, 1975; Keeling *et al.*, 1996). The observations at Mauna Loa and others only provide data for the last 40 years and data from pre-industrial times are needed as well. To do that the amount of carbon dioxide in air trapped in ice is measured. A spectacular rise in carbon dioxide can be seen and is some 25 per cent greater now then in pre-industrial times. Another notable feature of the recent observations of carbon dioxide is a slowing of the growth rate in 1992. Nearly all the greenhouse gases featured in this section also reveal a slowdown at this time too. The Mt

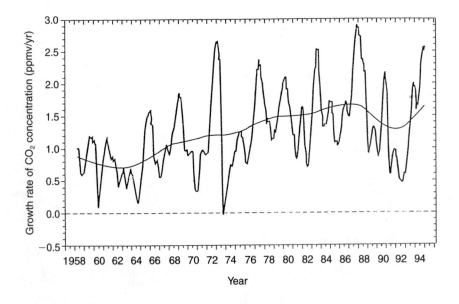

Figure 5.2 The growth rate of carbon dioxide concentrations as measured at Mauna Loa, Hawaii in ppmv/yr
Note: The smooth curve is the trend in concentration once variations on timescales less than 10 years have been removed
Source: Schimel *et al.* (1996)

Pinatubo eruption in 1991 may have been the cause. It has been suggested that the volcanic fallout included iron which acted as a fertilizer for oceanic phytoplankton. Carbon dioxide was then taken up by the increased numbers of phytoplankton (Whyte, 1995). What is clear from Figure 5.2 is that the growth rate of carbon dioxide is highly variable from year to year and over the decadal time-scale.

The carbon dioxide cycle together with the natural sources and sinks for carbon dioxide have already been discussed in Chapter 2. The amount of carbon dioxide release is measured by the amount of change of carbon in the reservoir. The units for this are gigatonnes of carbon per year or GtC/yr. The main anthropogenic source of carbon dioxide is fossil fuel burning. In addition, some industrial processes, such as cement production, release carbon dioxide directly into the atmosphere but their contribution is small. These sources released on average 5.5 ± 0.5 GtC/yr into the atmosphere during the 1980s (Schimel *et al.*, 1996). The other main source of carbon dioxide is land use changes and in particular deforestation. Uncultivated land and unmanaged forests can hold up to a 100 times more carbon than agricultural land (Watson *et al.*, 1990). Deforestation releases carbon dioxide into the atmosphere in a variety of ways: burning and decay of the biomass; oxidation of carbon held in the biomass and in the soil; and finally the renewal of the forest and soil after deforestation. Tropical deforestation and agriculture accounts for another 1.6 ± 1.0 GtC/yr being released into the atmosphere (Schimel *et al.*, 1996).

The first question to ask is, what happens to all the extra carbon dioxide being pumped into the atmosphere; does it just stay there? The amount of carbon dioxide in the atmosphere is far less than might be expected from known levels of anthropogenic carbon dioxide release. About half the carbon dioxide produced is taken up by the oceans and to a lesser degree by the biosphere. Anthropogenic production of carbon dioxide is estimated to be at a rate of 7 GtC/yr (Schimel *et al.*, 1996) and yet the atmospheric burden is only increasing by 3 GtC per year. Many believe that the remainder must be predominantly taken up in the oceans. At present the annual ocean uptake of carbon in the form of carbon dioxide is estimated to be 2 ± 0.8 GtC/yr. This uptake may have particular impact on global warming and the enhanced greenhouse effect. Changes in oceanic circulation and composition due to global warming will lead to changes in the oceanic transport of heat and greenhouse gases. This in turn will influence the ability of the oceans to take up carbon dioxide and impact further the climate system response to global warming. Therefore understanding atmosphere–ocean processes on a global scale is very important (Stanton, 1991).

Methane (CH₄)

Methane is produced by **anaerobic** processes which occur in a variety of settings. Natural wetlands such as bogs and swamps emit methane. Termites also emit methane in significant quantities. Both these natural sources can be directly influenced by human actions through land use changes. They may also be affected indirectly through climate change. Methane levels are positively correlated with global surface temperature on both the 10 000-year and shorter time-scales. This is due to climate feedback with climatic changes influencing the release of methane from wetlands (Houghton *et al.*, 1995). Methane is removed by chemical reactions in the atmosphere with the hydroxyl (OH) radical. A **radical** is an atom or, as in this case, a group of atoms with at least one unpaired electron. As a consequence the compound will not exist for long, reacting quickly with another to form a stable molecule. Therefore the main sink for methane occurs in the troposphere where other gases also compete to react with the hydroxyl radical. The removal of methane through this chemical reaction also results in the formation of carbon dioxide thereby producing an indirect contribution to global warming. There is also some removal of methane by the hydroxyl radical in the stratosphere. Soils may also remove methane from the atmosphere. However, because the main sink is a chemical reaction in the troposphere, then the lifetime of methane in the troposphere will depend on the reaction rate, i.e. just how fast that reaction occurs. It has been estimated that the lifetime of methane is somewhere in the region of 8 to 11.8 years, however, there is evidence to suggest that this lifetime is increasing because the amount of available hydroxyl is believed to be decreasing as the other gases which also react with it are increasing (Watson *et al.*, 1990).

As the human population has increased, so has the need to feed it. The increase in methane is an indirect result of this. When organic matter breaks down under anaerobic conditions methane is produced. Such conditions are associated with paddy rice cultivation and the digestive processes of cattle (Warrick *et al.*, 1990). Rice is predominantly grown in Asia and its production has doubled since 1940. The increase in sheep and cattle and the consequent gaseous emissions will have been offset somewhat by the decline in numbers of wild ruminants such as elephants and North American bison. However, methane emissions from this source are still believed to have risen significantly (Watson *et al.*, 1990). The organic waste that is piled into landfill sites is also thought to be a significant source of methane as it decays through anaerobic activity. Other human activities such as biomass burning, coal mining and natural gas production are all believed to add to methane emissions. It is thought that methane is now more than double it pre-industrial value of 0.8 ppmv at 1.72 ppmv (Watson *et al.*, 1990). The pattern of rising methane levels is somewhat difficult to interpret. In the 1970s the growth rate was around 20 parts per billion volume per year (ppbv/yr). This slowed in the 1980s to 13 ppbv/yr and by the mid-1990s was down to 8 ppbv/yr. In 1992 the growth of methane actually stopped at some locations. Why such an anomaly occurred in 1992/93 is unclear. Suggestions range from decreases in anthropogenic emissions to effects on the hydroxyl reaction in the atmosphere due to the volcanic eruption of Mt Pinatubo (Schimel *et al.*, 1996). Methane remains an important contributor to greenhouse gases because it is some 21 times more effective than carbon dioxide as a greenhouse gas. So although its atmospheric concentration is only 1 per cent that of carbon dioxide its contribution to global warming may be 17 to 20 per cent (Whyte, 1995).

Nitrous oxide (N₂O)

The emissions of nitrous oxide form part of the global nitrogen cycle (*see* Fig. 5.3). It is a complex cycle and consequently there are great uncertainties in the estimations of emissions from both natural and anthropogenic sources (Rosenzweig and Hillel, 1998). The main natural source for atmospheric nitrous oxide is thought to be soils. Most of the loss is from **aerobic** soils although significant contributions may be made under anaerobic conditions. The ocean is also a significant source of nitrous oxide. The main sink is in the stratosphere (Watson *et al.*, 1990). The nitrous oxide is broken up by the high energy short-wave sunlight that can penetrate into the stratosphere but not as far as the troposphere. As its destruction only takes place in the stratosphere, before being destroyed it has to move from the troposphere into the stratosphere, and as was explained in Chapter 2 the tropopause acts as a lid on the troposphere hindering transfers between the two. Consequently, nitrous oxide has a long lifetime in the atmosphere of 120 years. One molecule of nitrous oxide is 200 times more effective as a greenhouse gas compared to a molecule of carbon dioxide (Schimel *et al.*, 1996).

Anthropogenic sources of nitrous oxide are predominantly from agriculture and the use of fertilizers which grew significantly during the Second World War. A small but significant percentage of both nitrate and ammonia-based fertilizers can be converted into nitrous oxide. Further conversion may result from the **leaching** of fertilizers into groundwater. Nitrous oxide is now believed to be at 310 ppbv compared to its pre-industrial value of 287 ppbv and is increasing by a rate of 0.2 to 0.3 per cent per year (Whyte, 1995). Lower growth rates were observed in 1993 and again this has been attributed to the Mt Pinatubo eruption changing emission rates (Schimel *et al.*, 1996). There would have to be a reduction of 70 to 80 per cent of current emissions in order to stabilize the amount at present-day levels.

Halocarbons

Methyl chloride (CH₃Cl) is one of the few halocarbons that has a natural source, most are synthetic. It is produced by the oceans and by biomass

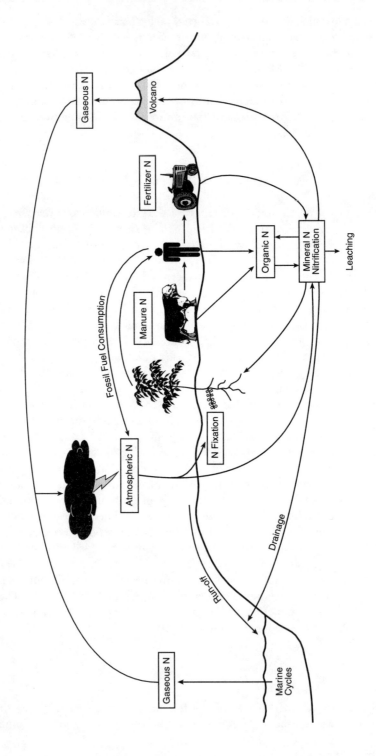

Figure 5.3 A simplified diagram of the nitrogen cycle

burning but it does not appear to be increasing (Watson *et al.*, 1990). The halocarbons that have received most attention of late are chlorofluorocarbons (CFCs). This group of chemicals are entirely synthetic, developed commercially in the early 1930s. Their original use was as a refrigerator liquid but they rapidly expanded into other areas. The reason for their widespread use is that they were seen as absolutely safe, non-reactive and non-toxic. Because they are so inert they do not react in the troposphere and consequently have long lifetimes. Like nitrous oxide they require the high energy sunlight that is to be found in the stratosphere to break them down. These photochemical reactions, together with other processes in the stratosphere, release free chlorine that can attack and deplete the ozone layer. This led to a serious thinning of the ozone layer at high latitudes and the 'ozone hole' over Antarctica (Farman *et al.*, 1985). The lifetime of CFCs varies considerably between the different types but the two predominant types, CFC-11 and CFC-12, have lifetimes of 65 and 130 years, respectively. Another group of halocarbons have a hydrogen atom, as well as carbon and halogens, as part of the compound. This group includes methyl chloride, Hydrochlorofluorocarbons (HCFCs) and Hydrofluorocarbons (HFCs). The latter two are being considered as replacements for CFCs in air conditioning and refrigeration. These are removed in the troposphere by reacting with the hydroxyl molecule, consequently they have shorter lifetimes than CFCs, ranging between 1 and 40 years.

CFC-11 and CFC-12 have concentrations of 280 parts per trillion volume (pptv) and 484 pptv respectively and they had been rising by at least 4 per cent per year. The international agreement signed in 1987, the **Montreal Protocol** and its Amendments, to protect the ozone layer by limiting ozone-depleting substances has already resulted in a significant slowing in the growth rates of CFCs. By 1994 CFC-12 levels had ceased to rise and CFC-11 concentrations were falling (Schimel *et al.*, 1996). It is estimated, however, that the atmospheric concentration of CFC-11 and CFC-12 will still be about 30 to 40 per cent of their current value for at least 100 years because of their long atmospheric lifetimes. All the halocarbons are highly efficient greenhouse gases, being far more effective than carbon dioxide. Even HCFCs and HFCs, which although less effective than CFCs, are still good greenhouse gases. Both HCFC and HFC atmospheric levels look set to grow as they replace CFCs in refrigerators and air conditioning. In the long term HCFCs may decrease, as they too deplete the ozone layer and will be phased out under the Montreal Protocol. In contrast HFCs do not deplete the ozone layer and their levels seem likely to be determined by individual countries' response to current and future greenhouse gas legislation. At present the amount of HFCs is very small and has little impact on global warming. However, IPCC projections estimate that by 2050 HFC emissions could have the same global warming impact as the current fossil fuel impact of France, Germany, Italy and the United Kingdom combined (French, 1997).

Tropospheric ozone

Stratospheric ozone transported into the troposphere is one natural source of tropospheric ozone. Tropospheric ozone is also formed by reactions between various other gases. These gases are nitrogen oxide (NO), nitrogen dioxide (NO_2), carbon monoxide (CO), methane (CH_4) and non-methane hydrocarbons (NMHC). The problem is that there is no simple formula to relate the amount of these precursor gases to the amount of tropospheric ozone that will be formed from them (Watson *et al.*, 1990). Not all these precursor gases are greenhouse gases themselves but we need to be aware of increasing anthropogenic sources of them because of the implications for changing the amount of tropospheric ozone. Carbon monoxide is formed due to incomplete combustion of fossil fuels and biomass and also from the oxidation of both natural and anthropogenic sources of methane and NMHC. It is predominantly removed from the troposphere by reacting with hydroxyl. Nitrogen oxides and dioxides are formed from fossil fuel and biomass burning and naturally by microbes in soils and by lightening. They are removed by conversion to nitric acid (HNO_3) and reactions involving the nitrate radical (NO_3). Methane sources and sinks have already been dealt with. NMHC's predominant source is the ocean although there are some anthropogenic sources such as biomass and fossil fuel burning. The major sink for NMHC is through reaction with the hydroxyl radical. The lifetime of tropospheric ozone is very short – approximately a few weeks. The same is true of several of the gas precursors. Due to its short lifetime tropospheric ozone is unevenly distributed across latitude, longitude, altitude and seasons. In the troposphere, below 8 km, it appears to be increasing at the rate of 1 per cent per year in the northern hemisphere (Watson *et al.*, 1990). However, globally in the 1980s there was little change in tropospheric ozone (Prather *et al.*, 1996).

Aerosols

Aerosols are not greenhouse gases but are included in this section to emphasize their importance to the radiation balance of the Earth. The effect of volcanic aerosols in the stratosphere has already been discussed in Chapter 4 and will not be further discussed here. Tropospheric aerosols have either a direct or indirect effect on the radiation balance. The direct effect is principally through the absorption or reflection of solar radiation. The indirect effect is through the modification of cloud optical properties and their lifetimes. Both of these effects depend heavily on the aerosol's size and chemical composition. Natural and anthropogenic aerosols form from primary and secondary sources (Jonas *et al.*, 1995). Natural aerosols are volcanic dust, soil dust and sea salt. These are fairly large aerosols, typically having a diameter of over 1 μm. In contrast, anthropogenic aerosols tend to be smaller and come from industrial dust, soot and biomass burning. Aerosols are quickly removed from the troposphere so they have very short atmospheric lifetimes. Consequently their spatial and temporal distribution is very varied. The net effect of anthropogenic aerosols is one of cooling. Sulphates released from fossil fuel burning and organic aerosols from biomass burning act like synthetic volcanoes reflecting away incoming radiation. However, soot probably has a small warming effect (Jonas *et al.*, 1995).

Global warming potential

The concept of a radiative forcing was introduced in Chapter 2. This is much easier to assess than the climatic response because of the feedbacks within the climate system. Radiative forcing can provide a simple measure of how important each greenhouse gas is relative to the others although the concept of radiative forcing has limitations when considering the full implications of climate change. Therefore there has been great emphasis in the various IPCC reports on determining the radiative forcing of each of the greenhouse gases. However, determining a useful indicator of radiative forcing has proved difficult. One way is the relative molecular forcing, the relative forcing on a molecule per molecule basis of each gas quoted relative to the carbon dioxide molecule. The forcing will depend on the change of concentration of each gas, but this is frequently not a linear relationship. Changes in the troposphere will result in changes in the stratosphere and consequently a change in radiative forcing at the tropopause. The amount of radiative forcing will alter depending on whether you wait until the stratospheric readjustment has occurred or whether you take an instantaneous value. Some molecules have both a direct and an indirect radiative forcing and both need to be taken into account. These include CFCs, which are a greenhouse gas so have a direct radiative warming effect, but also destroy the ozone layer which has an indirect cooling effect.

Global warming potential (GWP) is defined as the cumulative radiative forcing between the present and some later time 'horizon' caused by a unit mass of gas emitted now, expressed relative to some reference gas (Schimel *et al.*, 1996). GWP doesn't just take into account the change in concentration but also the lifetime of a gas, the longer the lifetime the longer it is around to change the radiative forcing. This is seen as a measure of emissions. However, as pointed out by Schimel *et al.* (1996), the GWP is far from the simple and straightforward measure that it is sometimes portrayed as. Some gases have both an indirect and direct effect on the radiative forcing of the Earth–atmosphere system. Including just one of the effects will give you a different GWP to using both. The GWP is usually only applied to a gas or aerosol that is well mixed in the atmosphere. Assumptions have to be made regarding the time horizons used and the reference gas. To illustrate some of these points consider Table 5.2. CFC-11 is a greenhouse gas. From just its direct effect, the GWP is 3800. However, CFC-11 also depletes the ozone layer: with less ozone trapping the ultraviolet radiation, the stratosphere cools. So less heat is radiated out by the stratosphere and the troposphere cools. This is the indirect cooling effect of CFC-11. Including this indirect effect as well as the direct one in the GWP calculation lowers the GWP of CFC-11 to 1300. HFCs, developed as a replacement for CFCs, can have very different GWPs depending on the reference gas used. Using carbon dioxide as a reference gas for HFC-134a gives a GWP of 1200 (1300 in Schimel *et al.*, 1996), however, using CFC-11 gives a GWP of 0.29. Climate change studies typically use carbon dioxide as the reference gas in GWP. The GWP given in Table 5.2 for HFC-134a is for a 100-year time horizon, for a 20-year time horizon the GWP is 3400 (Schimel *et al.*, 1996). Therefore great care should be used when trying to interpret claims about global warming based on GWPs.

Table 5.2 Global Warming Potentials (GWP) of selected halocarbons

Coolant	GWP given CO_2=1 100 yrs direct radiative effects only	GWP given CO_2=1 100 yrs@ indirect and direct radiative effects	GWP given R11=1
R11 (CFC)	3800@	1300–1700	1#
R12 (CFC)	8100@	6600–6800	3.1#
R22 (HCFC)	1600* 1500@	1300	0.37#
R123(HCFC)	93* 90@	30–50	
R411B(HCFC blend)	1000*		
R134a (HFC)	1200*	1300	0.29#
R404A(HFC blend)	3500*		

Sources:
*DTI (1995) direct radiative effects only
#DOE (1992)
@Schimel *et al.* (1996) ranges based on both indirect and direct radiative effects

Carbon dioxide observational data

In the previous chapter when discussing climate and its influence on history, emphasis was placed on determining the cause and effect. In particular these studies challenge the misuse of coincidence as being the evidence for cause and effect. Just because incident 'b' happened at a time when 'a' was occurring does not mean that 'a' caused 'b'. A similar rigorous approach should be used when considering the case of global warming. The courtroom analogy is often used at this point. More than circumstantial evidence is needed. Although a confession would be nice we are unlikely to get that. Instead we have to build a prosecution case that can withstand all that the defence can throw at it. This section will consider the most commonly asked questions about the observational record of carbon dioxide, for without a provable anthropogenic rise in carbon dioxide the prosecution case will crumble.

Natural variations

Most of Chapter 3 concentrated on how the amount of carbon dioxide in the atmosphere has changed continuously over geological time. How do we know that the current increase in carbon dioxide isn't just a natural variation? Watson *et al.* (1990) cite three pieces of evidence for this. First, the amount of carbon dioxide in air bubbles trapped in ice cores over the past thousand years has only varied slightly. That is about ten times less than the variation observed in the past 150 years. From the ice core record for the past 160 000 years, only during the last inter-glacial did carbon dioxide levels rise significantly to about 300 ppmv, which is still less than the 1990 value of 350 ppmv. Second, the rate at which carbon dioxide has increased in the atmosphere is extremely similar to the rise in the combined emissions from fossil fuel and land use changes. Although the rate of increase is similar, the increased amount of carbon dioxide in the atmosphere each year has always been less than that emitted by anthropogenic sources each year. From this scientists have inferred that the oceans and biota must be taking up this discrepancy, acting as sinks rather than sources of carbon dioxide. How long the oceans and biota can act in this way is unknown. Observations of carbon dioxide also show a south–north gradient in that there is more in the northern than the southern hemisphere. This gap has increased in line with the growth of northern hemisphere fossil fuel combustion. Third, by looking at the ratios of different carbon isotopes, scientists have concluded that these are consistent with those expected from anthropogenic emissions. It should be remembered that, if the global warming theory is right, any increase of carbon dioxide levels above the natural level will have committed the Earth to some sort of climate change. The fact that no distinct climate change has been observed so far does not mean that by keeping carbon dioxide levels at present day values nothing else will happen. The climate system, in particular the oceans, has not yet fully responded to the increase in carbon dioxide levels that has already taken place.

Following the IPCC 1990 report (Houghton *et al.*, 1990) there were suggestions that the variation in carbon dioxide was natural (Starr, 1993). This is based on the idea that the lifetime of carbon dioxide in the atmosphere is far shorter, about 10 years, than the widely accepted 100 years. If the lifetime of carbon dioxide is short, then it is quickly removed from the atmosphere. The large build-up of carbon dioxide that is currently being witnessed in the atmosphere could not then be due solely to anthropogenic emissions. These emissions would not be big enough when the quick removal rate was incorporated. Other natural emissions would have to be responsible for increasing carbon dioxide in the atmosphere. The short lifetime of carbon dioxide was arrived at by looking at carbon-14 released into the atmosphere after the atomic bomb tests in the 1960s. The 1995 IPCC report (Schimel *et al.*, 1996) quickly dismissed these claims pointing out that the uptake of carbon-14 by plants and the oceans cannot be used simply to compare to the uptake of excess carbon dioxide.

Absorption of long-wave radiation

Chapter 2 showed how absorption is spectrally selective. A molecule of a substance only absorbs certain wavelengths. If there is enough of that substance it will absorb all the radiation at that wavelength. In the case of the atmosphere such a gas would make the atmosphere opaque in that wavelength. There have been objections to global warming on the grounds that the carbon dioxide 15 μm absorption band is already saturated and so therefore increasing carbon dioxide will lead to no change in the amount of infra-red radiation trapped. However, looking at the absorption spectra of carbon dioxide as shown in Figure 1.10 reveals that there are many important absorption bands between 14 μm and 18 μm and these are far from saturated. It is at these wavelengths, above and below 15 μm, that absorption will take place and the greenhouse effect may be enhanced (Houghton *et al.*, 1990). The idea of the atmosphere acting like a greenhouse originates in Victorian times and carbon dioxide was identified as the main atmospheric constituent for controlling the global climate (Mudge, 1997). The idea was abandoned because it was thought that the absorption bands of water vapour would overlap with those of carbon dioxide. Therefore water vapour would absorb most of the radiation. Changes in carbon dioxide levels would then have little effect on the amount of radiation absorbed. By the mid-twentieth century this view had been reversed and carbon dioxide is again widely accepted as the predominant controller of global temperature (Mudge, 1997). Sceptics, however, still argue that the level of warming predicted by climate modellers is too high because of a neglect of the overlap between carbon dioxide and water vapour absorption (Idso, 1987; Potter *et al.*, 1987). Controversy also exists over the exact role that water vapour plays. The complex relationship between atmospheric composition and radiative profiles forms the basis of a major criticism of model predictions which will be discussed in Chapter 6.

Carbon dioxide fertilization

Plants can be classified into two different types called C3 and C4. This denotes the way in which they use carbon dioxide. For most plants current levels of carbon dioxide are too low for optimal photosynthesis. These are the C3 plants such as potatoes and barley, which in the EU, for example, account for over 90 per cent of the agricultural production (Warrick *et al.*, 1990). On a global scale C3 plants J168 include rice, bananas, coconuts and many other fruits and vegetables. Sixteen of the 20 most important food crops are C3 plants (Whyte, 1995). These plants will respond positively to a rise in carbon dioxide levels in the atmosphere leading to increased yields. This response is called the **carbon dioxide fertilization** effect. C4 plants such as maize will benefit little from increased carbon dioxide levels. One way in which both C3 and C4 plants will benefit is from reduced transpiration. The plant's water efficiency is increased which in turn increases yields. Under controlled trials, when crops are grown in an atmosphere with doubled carbon dioxide, all crop yields are increased. This has led some people, notably Idso (1991a, 1991b and 1998), to argue that carbon dioxide increases should be welcomed rather than seen as a catastrophe. Many people perceive the human population growth as the main environmental problem to worry about. In particular the ability (or lack of it) to feed that population has been a focus of concern for many centuries. Malthus writing in the eighteenth century was worried about the growth in population of the new urban poor in the rapidly industrializing European countries (Malthus, 1798). More recent work such as *Limits to Growth* considered other environmental threats such as resource depletion and pollution (Meadows and Meadows, 1972). Both works have as their main thrust that the Earth is reaching its carrying capacity and cannot cope with further increases in the human population. The argument for seeing global warming as a good thing is that increased yields will enable us to feed the ever-increasing human population. These views are, yet again, not new. Arrhenious in 1908 thought that increasing carbon dioxide would lead to a new 'carboniferous age' supporting the growing numbers of humans (Fleming, 1998). The trials showing increased yields with increasing atmospheric carbon dioxide, however, are carefully controlled. The plants are usually maintained in plastic bubbles or greenhouses where the ambient carbon dioxide levels can be strictly monitored. It is unclear what would happen naturally in a warmer world. For example pests such as Colorado Beetle on potatoes, whose distribution is limited by temperature, could become more widespread leading to a decrease in crop yields. Plants may adapt to the increasing carbon dioxide in the atmosphere and increased yields could be short-lived. There are also fears that the nutritional content of food grown under increased carbon dioxide levels will be reduced. Other laboratory experiments suggest that the benefits of increased carbon dioxide could be limited and highly dependent on water and nutrient availability (Bazzaz and Fajer, 1992).

Global warming or natural variability?

Perhaps the most contentious scientific statement made by the 1995 IPCC policy-makers' summary was that 'the balance of evidence suggests that there is a discernible human influence on global climate' (Houghton *et al.*, 1996). Figure 5.4 shows the global surface air temperature trend for the past 140 years. It reveals many variations and trends not just over hundreds of years but decades. Undoubtedly global temperatures have risen and the warmest years in the entire series occur between 1987 and 1996. Is this rise significant, however? Is it greater than the natural variability, or **noise**, of the climate system and does the rise have a natural or human origin? These are essentially the two problems facing climatologists: **'detection'** and **'attribution'**.

Detection means having to show that there has been a statistically significant change in climate. Attribution means that the cause has to be identified. Using the observed temperature trend to detect and attribute climate change is very difficult.

Researchers are looking to detect a signal. That signal is the predicted transient change in climate due to the increasing greenhouse gases. However, that will not be the only variation in the signal. There will be other variations not caused by growing greenhouse gases and these are called noise. Imagine you are on a stage at the front of a school hall crowded with children, all of whom have been assigned an individual number which they shout out. Your task is to point out the child shouting 46. That child is the signal and all the other children are the noise. You are not allowed to move about the children but child 46 has a microphone attached to a loud speaker. The longer the shouting goes on, the more the loudspeaker volume goes up until eventually you can hear it above all the other children. In essence that is what climatologists are trying to do, to detect the signal above the temperature noise and hoping to catch it as early as possible. Unfortunately things are not as straightforward for climatologists. Having identified the signal they have to be sure that it is really due to increased greenhouse gases and not to something else (attribution). In the analogy the signal would get so loud that you would not miss it. It is similar for the climate system. Global warming may become so widespread that it would be impossible for anything else other

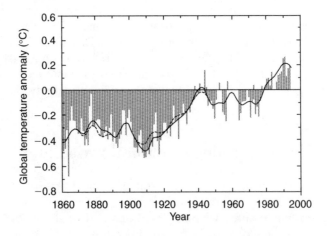

Figure 5.4 The change in global surface temperature (°C) from 1861 to 1994
Note: The temperature anomaly is found from comparing combined land-surface air and sea surface temperatures to a baseline temperature of 1961 to 1990. The bar chart represents annual values and the solid smooth curve the trend with variations on timescales less than ten years removed. The dashed curve is the same line as shown in 1992 IPCC report
Source: Houghton *et al.* (1996)

than increased carbon dioxide emissions to have caused it. However, it would seem prudent to detect global warming, if we can, before that stage.

In order to detect and attribute climate change to enhanced greenhouse gases, the climatologist first has to pick variables with good signal to noise ratios, that is the signal is strong in comparison to the noise. Going back to our school hall, imagine all the children shouting a number ending in 6. It would be much more difficult to detect child 46 early on, since the signal to noise ratio would be poor. Second, the climatologist has to decide how many variables to look at. The time series of a single variable can be used, the average global surface temperature for instance. But there are problems associated with it, as will be shown later. The preferred technique is the fingerprint method (Santer *et al.*, 1996). This method uses a single variable at many different locations, many variables at one location or several variables and locations. As an example of how this works, consider a single variable, temperature, but this time in three dimensions. A computer-based climate model is run to simulate the climate with increasing carbon dioxide used to provide the signal. These modelled fields are then compared to the observed. If there is no difference between the fields, then global warming has been detected, however, if there is a difference, then global warming has not yet occurred. The difference between the fields can be interpreted in several ways. It could be due to the signal being swamped by natural variability or that the observed or model fields are wrong. At the moment neither the observations or the computer model results can be said to be without problems. An improvement in the model results led to the 1995 IPCC statement that global warming had arrived (Hasselmann, 1997). However, there is the possibility that the agreement between the models and the observed data is merely fortuitous (Carson, 1999).

The observed temperature trend

Since the turn of the century the average global surface temperature has risen by just 0.5°C. This rise is well within the climate's natural variability or noise. To be really certain that the climate has changed, the enhanced greenhouse warming would have to rise above the noise. The temperature trend reveals lots of variation that cannot be attributed to increased carbon dioxide levels. For example, warming occurred between 1900–1940 when little should be expected and in the post-war period the northern hemisphere cooled. Therefore there are other natural effects as well as global warming occurring. From the temperature trend the natural variability has to be teased out to reveal any anthropogenic change. The temperature record shown in Figure 5.4 only covers the instrumental period. To fully explore the question of whether global warming has occurred a longer-term time series is required, around 1000 years. This requires that the temperature series is built up from proxy data and that brings with it all the problems discussed in earlier chapters. However, accepting that data we know that the climate has changed significantly in the past. Some have interpreted the recent warming trend as the maintenance of a recovery from the Little Ice Age and therefore a natural phenomenon. As pointed out by Nicholls *et al.* (1996), however, in some areas of the globe temperatures in the twentieth century have been the highest for thousands of years. There are also those who have suggested that global warming is required to save the world from a new Little Ice Age. The Marshall Institute (1989) argued that another Little Ice Age was imminent and that the expected cooling would offset most of the anthropogenically forced greenhouse warming. This is heavily contested and most scientists believe that global warming would be far greater than any global cooling that may occur over the next two centuries (Houghton, 1994; NRC, 1994; Shine *et al.*, 1990). Hulme and Jones (1994) consider the natural and other anthropogenic forcings that could cause the observed temperature trend. They conclude that the observed warming cannot be fully accounted for by these other factors.

The influence of the solar cycle on the Earth's climate has been subjected to a great deal of scrutiny since the discovery that the solar output varies (Parker, 1999). Friis-Christensen and Lassen (1991) used the length of the solar cycle as a proxy for solar irradiance. They concluded that most of the warming in the observed temperature trend could be accounted for by variations in the solar output. This has led to a long-running debate between those that support these findings and therefore argue that greenhouse gases have had little influence on the observed warming trend and those who refute the idea (Laut and Gundermann, 1998). This is particularly pertinent for the Sun seems to have entered an active phase like that of the medieval solar maximum (1100 to 1250) which is associated with the warm climate of the Medieval Optimum (Damon and Peristykh, 1999). Lockwood *et al.* (1999) looked at changes in the Sun's and Earth's magnetic field to calculate the Sun's energy output over the past 100 years. They concluded that up to half the global warming so far observed may be due to changes in the solar output. This is twice the figure widely accepted and would seem to add weight to the sceptics' argument, but their conclusions are far more complex than this. While prior to 1930 most of the global warming can be attributed to increased solar output, from 1970 only a third of the warming comes from the Sun. Lockwood *et al.* (1999) conclude that the remainder is from greenhouse gases and thus anthropogenic climate change is dominating. Several studies appear to echo this finding that the solar activity does have a significant influence on climate but that it cannot explain the entire observed surface warming (Damon and Peristykh, 1999; Kelly and Wigley, 1992; Laut and Gundermann, 1998).

Lane *et al.* (1994) use statistical analysis to try to separate out the global warming signal from all the other variations. They found that during the twentieth century carbon dioxide concentration was highly correlated with the temperature anomalies data. They point out that as this is a statistical analysis no cause and effect can be attributed. Also although the observed temperature trend is 0.45°C per century there are points in the record where there are decreases or no trend at all. Many of the criticisms of interpreting the upward trend of observed global surface temperatures have been based on the statistics used to analyse the trend. The simplest example of how the upward trend in the temperature data has been contested is in considering the baseline temperature. The baseline temperature is the average temperature for a chosen group of years. The temperature trend is then the year-on-year difference between a particular year's temperature and that of the baseline temperature. Therefore the anomaly calculated is going to depend critically on the group of years used to form the baseline. Sceptics have argued that the baseline is chosen to emphasize the warming trend. For example using the baseline 1961 to 1990 gives the average temperature over that period as 9.48°C. The 1991 to 1998 anomaly is +0.49°C. However, between 1931 and 1960 the average temperature was 9.60 °C and so the 1991 to 1998 anomaly compared to this baseline falls to +0.37 °C.

Another important feature of the observed temperature trend is when the warming is occurring. In the past few decades the difference between the minimum night-time and the daytime maximum temperature appears to have decreased in many regions of the globe. A substantial proportion of this decrease can be attributed to increasing night-time temperatures. Therefore the rise in mean observed temperature is mostly due to increasing night-time minimum temperatures (Mitchell *et al.*, 1995). The rise in minimum temperatures occurs in both rural and urban situations (Horton, 1995). It has been mainly attributed to an increase in cloud cover and tropospheric aerosol content along with modifications in soil moisture (Hansen, *et al.*, 1995). While the average temperature is important, it is often changes in the extreme temperatures, the daily minimum and maximum, that have the greatest implications for vegetation and health (Lockwood, 1998).

Reliability of the observational record

Although many of the problems with meteorological measurements were dealt with earlier, it is worthwhile summarizing them here and to mention some other criticisms that can be made of the observational record. A major problem is the incomplete coverage of the globe which makes it difficult to form a representative time series. Many changes in observing practice have occurred over time. Nowadays the positioning of a thermometer to obtain the best exposure is governed by World Meteorological Organization guidelines. However, these guidelines have changed and different countries may not have always followed the guidelines of the time. Other observing practices, in particular the time of the observation, may also change. This will alter the temperature series. There is no internationally accepted way for calculating the mean daily temperature, each country has its own formula. Although this may not affect the general trend of temperatures for that country, any change in the formula will distort the temperature series. Stations have changed location which brings a discontinuity to the temperature series. The instruments themselves have changed, causing inconsistencies in the data record. In calculating temperature trends climatologists make vigorous efforts to eliminate or acknowledge the errors that these effects might have on the measurements (Hulme and Jones, 1994).

One of the most contested systematic errors which may occur in the temperature record is the urban heat island effect. The **urban heat island** effect is well documented in microclimatology text books (Oke, 1990). Towns affect the atmospheric composition by introducing more pollutants than the rural surroundings but they also influence the heat budget. In general urban areas have higher temperatures than the surrounding countryside (*see* Fig. 5.5). Temperature differences between urban and rural locations are highest at night on calm, clear days and can be as much as 6°C. Meteorological observing stations are sometimes based in urban areas. As towns have expanded observing stations that were once rural are now in urban locations. Thus it has been suggested that the observed warming trend is due to the urban heat island effect not global warming (Balling and Idso, 1989). Estimates as to the importance of the urban heat island effect vary. Some estimates put it as low as 0.05°C per century and the IPCC 1990 report suggests a figure of 0.1°C per century (Folland *et al.*, 1990). Recent work suggests that the urban heat island effect may be larger than these estimates and that its full influence may not have been removed from the observed temperature series (Changnon, 1999). Overall the uncertainties in the global surface air temperature from land and sea surface observed data have been put by IPCC in 1995 as 0.15°C which results in a warming of between 0.3 and 0.6°C since the late 1800s (Nicholls *et al.*, 1996).

Due to the unreliability of surface data some researchers have looked to satellite data to provide a more objective estimate of changes in temperature.

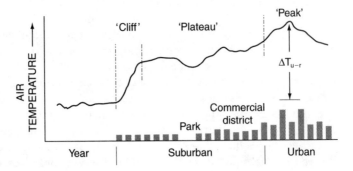

Figure 5.5 The urban heat island
Source: Oke (1990)

The tropospheric temperature trend from satellites has revealed a cooling of the order of 0.06°C per decade (Nicholls *et al.*, 1996). Such a result has provided those who see no threat from increased greenhouse gases with ample ammunition to challenge the surface-based observations (Pearce, 1997). However, these results should be looked at and interpreted with care. Satellites and their instrumentation are subject to changes, like surface data. Also satellite and surface measurements may not always be directly comparable, as they do not measure the same thing (Pearce, 1997). Most importantly satellite observations have only been available since the 1970s. The temperature series is therefore short and more subject to biases caused by the effects of volcanoes and the El Niño. Calculations have shown that when such biases are removed, the tropospheric temperature trends revealed by the satellites are more consistent with surface data with a warming of around 0.09°C per decade (Nicholls *et al.*, 1996). The results from satellites and radiosondes are more complex than surface measurements because they measure the temperature at different heights throughout the troposphere and stratosphere. As already mentioned, different parts of the atmosphere will respond to increased greenhouse gases in different ways. The lower troposphere has provided particularly contentious results with satellites showing a cooling of 0.05°C per decade since 1979 while surface measurements suggest a warming of 0.13°C per decade. Even the tenuous atmosphere at the height of the satellites' orbits creates a frictional drag on the satellite. This causes the satellite orbit to decay which means it very slowly falls to Earth. Recent research claims that this effect has never been fully considered when calculating the temperature of the lower troposphere from satellites. Once this is allowed for, satellite data shows a warming trend of 0.07°C per decade (Wentz and Schabel, 1998). At this point we could spiral into a debate about the applicability of such biases, their size, etc. However, let us just note that even the measurement of the temperature can be controversial.

Changes in other climatic variables

So much is said about temperature that there is a tendency to forget about the other parameters. Part of the problem is the lack of any long-term data sets. Precipitation, which is probably the next most widely analysed variable after temperature, is an excellent example. Instrumental data, like temperature, only extend back for the past 100 to 150 years. There is no marine precipitation data set, unlike temperature, so data analyses are restricted to land-based measurements only. Furthermore, there are no published satellite precipitation records (Hulme and Jones, 1994). Various analyses of precipitation over land show that over the past few decades it has increased in the northern hemisphere mid-latitudes and over the entire southern hemisphere. It has decreased in the northern hemisphere sub-tropics. However, there is a large amount of spatial variability (Nicholls *et al.*, 1996). Although the cryosphere is a highly integral part of the climate system, measurements of the various parameters are limited. Mountain glaciers appear to have retreated world-wide and it is thought that this is likely to be due to changes in temperature rather than changes in any other climatic variable such as precipitation and cloudiness which also affect glacial advance and retreat. Sea-levels have increased by 10 to 25 cm this century but the rate of rise has not been increasing. Global warming would cause a rise in sea level. As the oceans warm they expand and the sea level will rise accordingly. Melting glaciers will cause sea levels to rise as will melting of the Greenland ice sheet. The melting of the polar ice caps is very difficult to assess. The extreme cold, especially over Antarctica, as well as the subsidence of air, means that very little precipitation falls over the ice caps. If there is a warming, then more precipitation will fall in the interior so the ice sheets will grow, leading to a sea level drop. It is thought that over Greenland the melting at the margins will more than offset the accumulation of ice in the interior. Over the colder Antarctica the reverse is thought to be true. However, there is great uncertainty about the reaction of the ice sheets to global warming and what changes have taken or will take place. A lot of concern has focused on the possibility of the rapid melting of the West Antarctic Ice Sheet leading to major sea level rise (Oppenheimer, 1998). The 1995 IPCC report felt this was unlikely to occur by the year 2100 (Warrick *et al.*, 1996). Oppenheimer (1998) agrees that any collapse of the West Antarctic Ice Sheet will probably occur after 2100. However he points out that the necessary greenhouse gas concentrations for such a collapse are likely to be reached during the twenty-first century.

Extreme weather events

This aspect of global warming is the one that concerns many people. Will events such as the heat wave that affected much of the USA in 1988 occur more often? Popular images of global warming are of increased extreme conditions and excessive sea level rise. It is these extreme events that would cause the most damage and may have the most impact. Mitchell *et al.* (1990) detail how the frequency of extreme events might change. Climatic events such as daily average temperature or rainfall form a bell-shaped or normal distribution. This is shown by the solid line in Figure 5.6. Taking temperature as our example, with global warming, mean temperature may increase. The shape of the curve remains unchanged but shifts right. This results in a larger number of hot days but fewer cold days (Figure 5.6a). Alternatively, the mean temperature may stay the same but the shape of the curve might change as shown in Figure 5.6b. In this case the frequency of hot and cold days will increase. Alternatively, the mean and the distribution may change together.

Many problems arise in the analysis in variability. How do you set a baseline, an average climate, from a constantly changing climate? How do you define what is extreme? Changes in meteorological practice and analysis in defining such extremes cause problems when trying to assess variability. As a result, many changes have been found to be due to these inconsistencies. Only **tropical cyclones** (Hurricanes) in the North Atlantic are thought to have changed and that was a decrease in activity (Nicholls *et al.*, 1996). There is some evidence for less extreme cold temperatures, but no increase in high temperature events, even though mean temperatures have increased. Given the lack of global data on the time-scales needed, no firm conclusions can yet be drawn, but there appears to be no increase in extreme weather events or climate variability at global scales. At regional scales there is some evidence of change, but the change is not uniform with some areas exhibiting greater variability and others lower variability.

One area where there have been distinct changes in variability is in the El Niño Southern Oscillation (ENSO) described in Chapter 2. Since 1976–77 the El Niño has returned far more frequently than previously documented. Instead of the normal 4 to 5 year gap between each El Niño the return period has been about 2 years (Nicholls *et al.*, 1996). There have been few La Niña episodes. As a result there are distinct changes in the precipitation and circulation patterns. This may explain the decrease in hurricane activity, noted earlier, as El Niños are associated with weakened hurricane activity. The El Niño of 1997 was the largest-amplitude event to have taken place since instrumental records began (Jones and Thorncroft, 1998). It would appear that the current ENSO behaviour has not occurred since the instrumental record began about 120 years ago. Whether the ENSO has ever exhibited behaviour similar to the present pattern prior to the instrumental record is unclear (Nicholls *et al.*, 1996). Such studies have to rely on historical documentation of an event, which may well be missing. They also require some means of assessing the

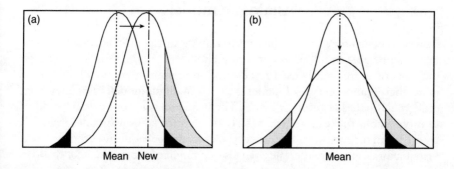

Figure 5.6 How changes in the frequency distribution can affect the number of extreme events: (a) by increasing the mean the number of extreme events on the right-hand side is increased, (b) the mean is unchanged but the standard deviation is increased (the amount of variability). In this case extreme events on both sides of the distribution occur
Source: Mitchell *et al.* (1990)

strength of the ENSO. This is a difficult process and depends heavily on the measure used. For example it is unclear whether the ENSO event of 1997–98 was truly more severe than the 1982–83 event (Wolter and Timlin, 1998). An analysis of the ENSO events since 1500 by Quinn and Neal (1992) concludes that there are long-term changes in activity lasting several decades. Furthermore, in comparison with the average activity over the past 500 years, the period between 1976 to 1987 was a time when a below average number of events occurred (Figure 5.7). This could be an artefact of the proxy data and conflicts with the conclusions of Nicholls *et al.* (1996). Duxbury and Duxbury (1989) focus on the lack of La Niña events during this period. They point out that these colder than average events tend to be ignored because the La Niña does not have such a significant socio-economic impact. This is despite the fact that the very strong La Niña in 1988 may have been a major cause of the droughts in central USA in 1988. ENSO events are associated with higher than normal sea surface temperatures in a large area of the Pacific Ocean. This has led some to speculate that it is the increase in El Niño events that has led to an increase in surface temperatures, rather than global warming leading to an increase in El Niños. It seems likely that ENSO will not only be affected by global warming, but will in turn, affect the climate system's response to global warming (Dickinson *et al.*, 1996; Duxbury and Duxbury, 1989). Changes in the North Atlantic Oscillation (NAO) have also been used to question whether global warming is due to carbon dioxide increase or natural changes in the atmosphere–ocean interactions (Morton, 1998). From 1990 until 1995 the NAO index was extremely high, leading to warmer European winters. While this resembles results found in doubled carbon dioxide experiments with climate models, it is unclear how changes in the NAO cycle have contributed to the warming (Hurrell, 1995).

Analogue models

As well as informing us about the current climate and its variability, observations may also be able to tell us about future climates. Although the term climate model is most frequently used to describe computer-based models of the climate, there is another type of model called an **analogue model**. Past climates are used as analogues of future climates. There are two types of analogue models. One type looks to the geological past, to times when carbon dioxide levels were greater than they are now. The other type of analogue model uses the historical instrumental record to infer the changes that might occur in a warmer world.

Paeleo-analogue models

Climatologists using this method to predict the consequences of global warming need to look for periods in the Earth's past when the carbon dioxide

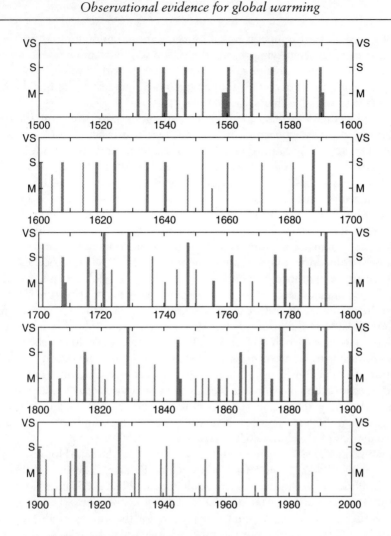

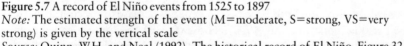

Figure 5.7 A record of El Niño events from 1525 to 1897
Note: The estimated strength of the event (M=moderate, S=strong, VS=very
strong) is given by the vertical scale
Source: Quinn, W.H. and Neal (1992). The historical record of El Niño, Figure 32.1.
In Bradley, R.S. and Jones, P.D. (eds), *Climate since A.D. 1500*. London: Routledge

levels were significantly above pre-industrial levels. There is strong geological
evidence that for several periods in the geological past the carbon dioxide levels in the atmosphere were far greater than they are now. Having identified
these periods scientists then need to derive from the geological data just how
much carbon dioxide was in the atmosphere. Then, using current predictions
of future emissions of carbon dioxide, they can suggest a date in the future
when the carbon dioxide level in the atmosphere will correspond to that in
the past. Having reached this stage geological data are then used to try and

reconstruct the past temperature and precipitation, building up a picture of the past climates. Achieving that will give the scientists an analogue for the climate in the future which has the same carbon dioxide levels as the one in the past. For a fuller account of analogue modelling details see Budyko and Izrael (1991) or Cubasch and Cess (1990).

The first stage in paeleoclimatic reconstructions is to determine the temperature sensitivity to carbon dioxide changes (Cusbasch and Cess, 1990). Budyko *et al.* (1987) obtained global mean temperature changes for four periods: Early Pliocene (7 to 6 million BP), early and middle Miocene (26 to 14 million BP), Paeleocene-Eocene (60 to 54 million BP) and Cretaceous (136 million BP). The temperature changes associated with these times have to be adjusted to allow for the differences in the solar constant and albedo in the geological past. The solar constant is assumed to increase by 5 per cent every billion years. A 1 per cent change corresponds to a $1.4°C$ rise in global mean temperatures. Changes in albedo are due to changes in the ratio of continents to oceans. Each 0.01 reduction in albedo is assumed to raise the mean global temperature by $2°C$. These changes account for 20 to 50 per cent of the temperature changes. Next the residual temperature change has to be related to the concentration of carbon dioxide in the atmosphere. Atmospheric carbon dioxide levels were estimated using a carbon cycle model. Budyko *et al.* (1987) found the Eocene concentrations to be five times greater than present and the Cretaceous concentrations to be nine times greater. Other workers (Shackleton, 1985) doubt atmospheric carbon dioxide was ever greater than twice today's value, the rest being taken in by the oceans. The result is a sensitivity of $3.0°C±1°C$ for doubling carbon dioxide. This result is similar to results found with computer-based models of the climate (Cusbasch and Cess, 1990).

Budyko *et al.* (1987) went on to use three periods: the mid-Holocene optimum (5000–6000 BP), the Last-Interglacial or Eemian (125 000–130 000 BP) and the Pliocene (3–4 million BP), as analogues for 2000, 2025 and 2050 respectively. January and July mean air temperatures and annual precipitation were reconstructed. The mean annual temperature for the three periods were 1, 2 and 3 to $4°C$ warmer than in the nineteenth century (Budyko *et al.*, 1987). There would also appear to have been a general increase in precipitation in all three periods (Folland *et al.*, 1990). The nature of the forcing causing these temperature changes in each of the periods was different. During the Pliocene the climate changes have been related to increased carbon dioxide levels, perhaps as much as twice pre-industrial levels. During the Eemian inter-glacial period carbon dioxide levels were only slightly raised and orbital variations may provide the remaining forcing required (Folland *et al.*, 1990). Despite these variations in forcing the relative temperature change over each hemisphere were similar. Budyko (1991) argues that this together with temperature anomaly maps shows that empirical methods for estimating the spatial distribution of temperature change to global warming may be relatively robust.

Pittock and Salinger (1982) review several modelling methods to produce regional scenarios for a carbon dioxide warmed world. They were particularly interested in the southern hemisphere. Using paeleo-analogue methods Pittock and Salinger (1982) used the climatic optimum of the Holocene. They consider 9000 to 7000 BP as the best dates for the maximum in Australasia, which is much earlier than in Europe and America. There is evidence for a much warmer and wetter regime than now, from lake levels, vegetation types and pollen. Pittock and Salinger (1982) point out that the main objection to this modelling is that although atmospheric carbon dioxide may have exceeded present values, the mid-Holocene warming was not caused by an increase of carbon dioxide. However, they argue that the approach is valid because the modelling shows very little difference in the warming effects shown by increased carbon dioxide or solar radiation.

Historical analogue technique

In this technique historical meteorological records of observations are used. A group of warm years and a group of cold years are extracted from the instrumental record and compared. The groups can be composites of individual cold and warm years or groups of consecutive cold and warm years. Differences in the climatic variables between the two groups of years are then associated with the temperature difference. There can be no attempt to associate cause and effect using analogue models. It relies on the associations between the observed values being representative of physical processes. So although we cannot say that an increase in temperature will cause an increase in cloud amount, we can say that an increase in temperature is associated with an increase in cloud amount. It is a subtle difference but a very important one.

Pittock and Salinger (1982) used data from the Antarctic and Southern Ocean data set, which at that time was only 23 years long. They used a composite of seven cool years (1958, 1959, 1963, 1964, 1965, 1969 and 1976) and nine warm years (1957, 1961, 1968, 1970, 1971, 1973, 1974, 1975 and 1977) and noted how warm and cool years tend to bunch together. Warm years were defined as being when above average temperatures occurred in the Antarctic and Southern Ocean regions or one was above and the other not below average temperature. There are 8 to 11 observing stations in each region. The mean difference between the warm and cool years were 0.9 K in Antarctica and 0.3 K for the Southern Ocean. The precipitation differences between these warm and cool years for Australia and New Zealand are shown in Figure 5.8. The whole of Australia north of 30°S is appreciably wetter in the warmer years. In New Zealand the temperature difference between the two periods is 0.5 K. Significant precipitation increases occur in the eastern areas of the island. The overall pattern is statistically significant for New Zealand but not Australia. Pittock and Salinger (1982) point out that this method may be only indicative of the transitory response to carbon dioxide. Even this may be in doubt, though Wigley *et al.* (1980) argue that the changes will be internally consistent. There are also reservations about how to define a warm year and what series to use.

Henderson-Sellers (1986a and b) has produced several studies looking at cloud amount and temperature in warm and cold periods. Understanding how clouds will respond to increased temperatures is important because of their strong effect on the Earth's radiation budget. The studies concentrated on regional cloud amounts, that is continental cloud amounts in the northern hemisphere during the period 1900 and 1954. Rather than using individual years a series of cold and warm years were intercompared. The warm period was defined as being the 20 years between 1944 to 1953 and the cold period was 1901 to 1920. Remember this is looking at transient temperature increases and also the study results will contain any signal due to changes in observing mode. There are probably other factors contributing to the observed temperature increase apart from carbon dioxide (about half). Henderson-Sellers (1986a and b) found that a small temperature rise of around 0.3°C was associated with a cloud amount increase of between 0.1 and 0.3 tenths.

Criticisms

The major criticism of analogue models is that the forcing mechanisms in the past may not mimic those of the future. Therefore there is no guarantee that the climate's response will be the same with different forcings. The problems that affect paeleo-analogue methods often relate back to the more general problem of reconstructing geological climates. These criticisms include the use of relatively local geological data to represent a large area; the imprecise dating of the proxy data, and disagreement over the reconstructed climate (Cusbasch and Cess, 1990).

Summary

Our ability to observe planet Earth has increased immeasurably in the last half of the twentieth century. This vast increase in data, however, would appear to bring with it as many questions as answers. This chapter has revealed the different stances taken by scientists as they try to comprehend these data. Some scientists view global warming as a positive event. It would seem fair to say that as yet the rise in surface temperature is not outside the natural variability of the climate system. However, in assessing all the possible contributory forcings coupled with the rate of rise in temperature, the most likely cause for this would seem to be increasing carbon dioxide levels. It will probably take several more decades before we can truly know whether global warming is happening. In that time, if there is no action to curb greenhouse gases, the levels will continue to rise and the

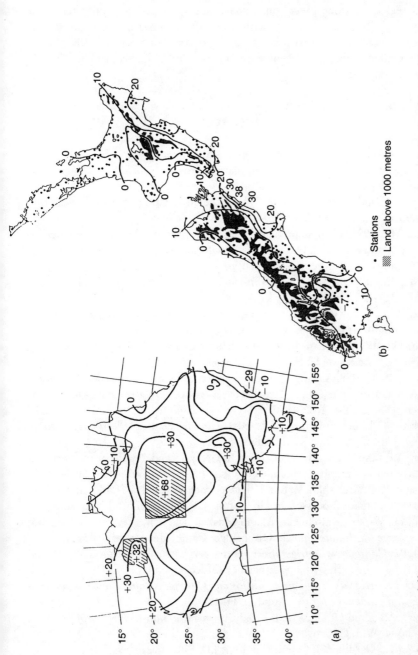

Figure 5.8 Percentage differences in mean annual precipitation between cool and warm years over (a) Australia (hatching indicates a statistically significant difference at the 90 per cent confidence interval) and (b) New Zealand (statistically significant) from an analogue model

Source: Pittock and Salinger (1982).

potential change in climate will become greater. This suggests that we should seek to reduce greenhouse gas levels now, rather than later, as a precaution. Chapters 7 and 8 will show that even this decision is not without difficulty or controversy. Chapter 6 will consider the other main type of evidence that is presented in the case of global warming, that of computer model data.

References

Balling, R.C. Jr and Idso, S.B. 1989: Historical temperature trends in the United States and the effect of urban population growth. *Journal of Geophysical Research* **94**, 3359–63.

Bazzaz, F.A. and Fajer, E.D. 1992: Plant life in a CO_2-rich world. *Scientific American*, January, 18–24.

Budyko, M.I. 1991: Climate prediction based on past and current analogues. In Jaegar, J. and Ferguson, H. L. (eds), *Climate change: science impacts and policy*. Cambridge: Cambridge University Press.

Budyko, M.I. and Izrael, Y. 1991: *Anthropogenic climate change*, English translation. Tucson: The University of Arizona Press.

Budyko, M.I., Ronav, A.B. and Yanshin, Y.L. 1987: *History of the Earth's atmosphere*. English translation. Berlin: Springer-Verlag.

Carson, D.J. 1999: Climate modelling: achievements and prospects. *Quarterly Journal of the Royal Meteorological Society* **125**, 1–27.

Changnon, S.A. 1999: A rare long record of deep soil temperatures defines temporal temperature changes and an urban heat island. *Climatic Change* **42**, 531–8.

Cubasch, U. and Cess, R.D. 1990: Processes and modelling. In Houghton, J.T., Jenkins, G.J. and Ephraums, J.J. (eds), *Climate change: the IPCC scientific assessment*. Cambridge: Cambridge University Press.

Damon, P.E. and Peristykh, A.N. 1999: Solar cycle and 20th century northern hemisphere warming: revisited. *Geophysical Research Letters* **26**, 2469–72.

Dickinson, R.E., Meleshko, V., Randall, D., Sarachik, E., Silva-Dias, P. and Slingo, A. 1996: Climate Processes. In Houghton, J.T., Meira Filho, L.G., Callander, B.A., Harris, N., Kattenberg, A. and Maskell, K. (eds), *Climate change 1995: the science of climate change*. Cambridge: Cambridge University Press.

DOE 1992: *Commercial refrigeration plant: energy efficient design*, no. 37 Good Practice Guide Series, Energy Efficiency Office. London: HMSO.

DTI 1995: *Refrigeration and air conditioning: CFC phase out: advice on alternatives and guidelines for use*. London: HMSO.

Duxbury, A.C. and Duxbury, A.B. 1989: *An introduction to the world's oceans*. Dubuque: Wm. C. Crown.

Farman, J.C., Gardiner, B.G. and Shanklin, J.D. 1985: Large losses of total ozone in Antarctica reveal seasonal ClO_x/NO_x interaction. *Nature* **315**, 207–10.

Fleming, J.R. 1998: Arrhenious and current climate concerns: continuity or 100-year gap?, *Eos* **79**, 405, 409, 410.

Folland, C.K., Karl, T.R. and Vinnikov, K. Ya. 1990: Observed climate variations and change. In Houghton, J.T., Jenkins, G.J. and Ephraums, J.J. (eds), *Climate change: the IPCC scientific assessment*. Cambridge: Cambridge University Press.

French, H. F. 1997: Learning from the Ozone Experience. In Starke, L. (ed.), *State of the World 1997*. New York: W.W. Norton and Company.

Friis-Christensen, E. and Lassen, K. 1991: Length of the solar cycle: an indicator of solar activity closely associated with climate. *Science* **254**, 698–700.

Hall, C.A.S., Ekdaht, C.A. and Wartenberg, D.E. 1975: A fifteen-year record of biotic metabolism in the Northern hemisphere. *Nature* **255**, 136–8.

Hansen, J., Sato, M. and Ruedy, R. 1995: Long-term changes of the diurnal temperature cycle: implications about mechanisms of global climate change. *Atmospheric Research* **37**, 175–209.

Hasselmann, K. 1997: Are we seeing global warming? *Science* **276**, 914–15.

Henderson-Sellers, A. 1986a: Cloud changes in a warmer Europe. *Climatic Change* **8**, 25–52.

Henderson-Sellers, A. 1986b: Increasing cloud in a warmer world. *Climatic Change* **9**, 267–309.

Horton, E.B. 1995: Geographical distribution of changes in maximum and minimum temperatures. *Atmospheric Research* **37**, 102–17.

Houghton, J. 1994: *Global warming: the complete briefing*. Oxford: Lion Publishing.

Houghton, J.T., Callander, B.A. and Varney, S. K. (eds) 1992: *Climate change 1992: the supplementary report to the IPCC scientific assessment*. Cambridge: Cambridge University Press.

Houghton, J.T., Jenkins, G.J. and Ephraums, J.J. (eds) 1990: *Climate change: the IPCC scientific assessment*. Cambridge: Cambridge University Press.

Houghton, J.T., Meira Filho, L.G., Bruce, J., Hoesung Lee, Callender, B.A., Harris, N. and Maskell, K. (eds) 1995: *Climate change 1994: radiative forcing of climate change and an evaluation of the IPCC IS92 emission scenarios*. Cambridge: Cambridge University Press.

Houghton, J.T., Meira Filho, L.G., Callander, B.A., Harris, N., Kattenberg A. and Maskell, K. (eds) 1996: *Climate change 1995: the science of climate change*. Cambridge: Cambridge University Press.

Hulme, M. and Jones, P.D. 1994: Global climate change in the instrumental period. *Environmental Pollution* **83**, 23–36.

Hurrell, J.W. 1995: Decadal trends in the North Atlantic Oscillation and its relation to regional temperature and precipitation. *Science* **269**, 676–9.

Idso, S.B. 1987: A clarification of my position on the CO_2-climate connection. *Climatic Change* **10**, 81–6

Idso, S.B. 1991a: Viewpoint: CO_2 – a blessing in disguise. *Weather* **46**, 220.

Idso, S.B. 1991b: Carbon dioxide and the fate of the Earth. *Global Environmental Change* **1**, 178–82.

Idso, S.B. 1998: CO_2-induced global warming: a skeptic's view of potential climate change. *Climate Research* **10**, 69–82.

Jonas, P.R., Charlson, R.J. and Rodhe, H. 1995: Aerosols. In Houghton, J.T., Meira Filho, L.G., Bruce, J., Hoesung Lee, Callender, B.A., Harris, N. and Maskell, K. (eds), *Climate change 1994: radiative forcing of climate change and an evaluation of the IPCC IS92 emission scenarios.* Cambridge: Cambridge University Press.

Jones, C.G. and Thorncroft, C.D. 1998: The role of El Niño in Atlantic tropical cyclone activity. *Weather* 53, 324–36.

Keeling, C.D., Chin, J.F.S. and Whorf, T.P. 1996: Increased activity of northern vegetation inferred from atmospheric CO_2 measurements. *Nature* 382, 146–9.

Kelly, P.M. and Wigley, T.M.L. 1992: Solar cycle length, greenhouse forcing and global climate. *Nature* 360, 328–30.

Lane, L.J., Nichols, M.H. and Osborn, H.B. 1994: Time series analyses of global change data. *Environmental Pollution* 83, 63–8.

Laut P. and Gundermann, J. 1998: Solar cycle hypothesis appears to support IPCC on global warming. *Journal of Atmospheric and Solar–Terrestrial Physics* 60, 1719–28.

Lockwood, J.G. 1998: Future trends in daytime and night-time temperatures. *Weather* 53, 72–8.

Lockwood, M., Stamper, R. and Wild, M.N. 1999: A doubling of the Sun's coronal magnetic field during the past 100 years. *Nature* 399, 437–9.

Malthus, T. 1798: *Essay on the principle of population* (reprinted 1926). London: Macmillan.

Marshall Institute, 1989: *Scientific perspectives on the greenhouse problem.* Seitz, F. (ed.), Washington, DC: Marshall Institute.

Meadows, D.H. and Meadows, D.L. 1972: *The limits to growth: a report to the Club of Rome's project on the predicament of mankind.* New York: Universe Books.

Mitchell, J.F.B., Davis, R.A., Ingram, W.J. and Senior, C.A. 1995: On surface temperature, greenhouse gases and aerosols: models and observations. *Journal of Climate* 10, 2364–86.

Mitchell, J.F.B., Manabe, S., Meleshko, V. and Tokioka, T. 1990: Equilibrium climate change – and its implications for the future. In Houghton, J.T., Jenkins, G.J. and Ephraums, J.J. (eds), *Climate change: the IPCC scientific assessment.* Cambridge: Cambridge University Press.

Morton, O. 1998: The storm in the machine. *New Scientist* 157 (no. 2119), 22–7.

Mudge, F.B. 1997: The development of the 'greenhouse' theory of global change from Victorian times. *Weather* 52, 13–7.

Nicholls, N., Gruza, G.V., Jouzel, J., Karl, T.R., Ogallo, L.A. and Parker, D.E. 1996: Observed climate variability and change. In Houghton, J.T., Meira Filho, L.G., Callander, B.A., Harris, N., Kattenberg, A. and Maskell, K. (eds), *Climate change 1995: the science of climate change.* Cambridge: Cambridge University Press.

NRC 1994: *Solar influences on global change.* Board on global change commission on geosciences, environment and resources, National Research Council. Washington, DC: National Academy Press.

Oke, T.R. 1990: *Boundary layer climates*. London: Routledge.

Oppenheimer, M. 1998: Global warming and the stability of the West Antactic Ice Sheet. *Nature* **393**, 325–32.

Parker, E.N. 1999: Solar physics – sunny side of global warming. *Nature* **399**, 416–7.

Peace, F. 1997: Greenhouse wars. *New Scientist* **155** (no. 2091), 38–43.

Pittock, A.B. and Salinger, M.J. 1982: Towards regional scenarios for a CO_2-warmed Earth. *Climatic Change* **4**, 23–40.

Potter, G.L., Kiehl, J.T. and Cess, R.D. 1987: A clarification of certain issues related to the CO_2-climate problem. *Climate Change* **10**, 87–95.

Prather, M., Derwent, R., Enhalt, D., Fraser, P., Sanhueza, E. and Zhou, X. 1996: Other trace gases and atmospheric chemistry. In Houghton, J.T., Meira Filho, L.G., Callander, B.A., Harris, N., Kattenberg, A. and Maskell, K. (eds), *Climate change 1995: the science of climate change*. Cambridge: Cambridge University Press.

Quinn, W.H. and Neal, V.T. 1992: The historical record of El Niño. In Bradley, R.S. and Jones, P.D. (eds), *Climate since AD 1500*. London: Routledge.

Rosenzweig, C. and Hillel, D. 1998: *Climate change and the global harvest*. Oxford: Oxford University Press.

Santer, B.D., Wigley, T.M.L., Barnett, T.P. and Anyamba, E. 1996: Detection of climate change and attribution of causes. In Houghton, J.T., Meira Filho, L.G., Callander, B.A., Harris, N., Kattenberg, A. and Maskell, K. (eds), *Climate change 1995: the science of climate change*. Cambridge: Cambridge University Press.

Schimel, D., Alves, D., Enting, I., Heinmann, M., Joos, F., Raynaud, D., Wigley, T., Prather, M., Derwent, R., Ehhalt, D., Fraser, P., Sanhueza, E., Zhou, X., Jonas, P., Charlson, R., Rodhe, H., Sadasivan, S., Shine, K.P., Fouquart, Y., Ramaswamy, V., Solomon, S., Srinivasan, J., Albitton, D., Derwent, R., Isaksen, I., Lal, M. and Wuebbles, D. 1996: Radiative forcing of climate change. In Houghton, J.T., Meira Filho, L.G., Callander, B.A., Harris, N., Kattenberg, A. and Maskell, K. (eds), *Climate change 1995: the science of climate change*. Cambridge: Cambridge University Press.

Shackleton, N.J. 1985: Oceanic carbon isotope constraints on oxygen and carbon dioxide in the Cenozoic atmosphere. 'The carbon cycle and atmospheric CO_2: natural variations, Archaean to present'. *Geophysical Monographs* **32**, 412–7.

Shine, K.P., Derwent, R.G., Weubbles, D.J. and Morcrette, J-J. 1990: Radiative forcing of climate. In Houghton, J.T., Jenkins, G.J. and Ephraums, J.J. (eds), *Climate change: the IPCC scientific assessment*. Cambridge: Cambridge University Press.

Stanton, H.R., 1991: Ocean circulation and ocean–atmosphere exchanges. *Climatic Change* **18**, 175–94.

Starr, C. 1993: Atmospheric CO_2 residence time and the carbon cycle. *Energy* **18**, 1297–310.

Warrick, R.A., Barrow, E.M. and Wigley, T.M.L. 1990: *The greenhouse effect and its implications for the European Community*. Belgium: Commission of European Communities.

Warrick, R.A., Le Provost, C., Meier, M.F., Oerlemans,. J. and Woodworth, P.L. 1996: Changes in sea level. In Houghton, J.T., Meira Filho, L.G., Callander, B.A., Harris, N., Kattenberg, A. and. Maskell, K. (eds), *Climate change 1995: the science of climate change*. Cambridge: Cambridge University Press.

Watson, R.T., Rodhe, H., Oeschger, H. and Siegenthaler, U. 1990: Greenhouse gases and aerosols. In Houghton, J.T., Jenkins, G.J. and Ephraums, J.J. (eds), *Climate change: the IPCC scientific assessment*. Cambridge: Cambridge University Press.

Wentz, F. J. and Schabel, M. 1998: Effects of orbital decay on satellite-derived lower tropospheric temperature trends. *Nature* 394, 661–4.

Whyte, I.D. 1995: *Climatic Change and Human Society*. London: Arnold.

Wigley, T.M.L., Jones, P.D. and Kelly, P.M. 1980: Scenario for a warm high-CO_2 world. *Nature* 283, 17–21.

Wolter, K. and Timlin, M.S. 1998: Measuring the strength of ENSO events: How does 1997/1998 rank? *Weather* 53, 315–24.

6

The model evidence for global warming

Introduction

The previous chapter showed that there is a lot of observational evidence which suggests that the amount of greenhouse gases in the atmosphere and the global average surface temperature are rising. By now though you should realize that linking the two is far from straightforward. It is almost an impossible task to prove, before any significant climate change, that it is human pollution that is responsible for both these upward trends. So apart from the physics and the observations, scientists have also used computer models of the climate in an attempt to understand what is happening and to provide further evidence about the potential climate change. These models provide the best chance of attributing climate change to increasing greenhouse gases.

The most commonly used models are vast computer programs which attempt to simulate the Earth's climate. As they can simulate a changed climate they can tell us about the future, about what our weather and climate will be like. So the models could help in formulating policy decisions about what humans should do about global warming. This chapter will describe the various models that are available to climatologists as well as giving a summary of the current model predictions. It will also look at the various criticisms that have been levelled at the models and why some scientists feel that as yet the models are inadequate for informing policy. It will be left until the final chapter to fully discuss where that leaves policy-makers.

Computer-based models

Development

The most widely used models are computer-based climate models. These models attempt to predict the climate from some basic physical principles.

This means that the equations which govern the climate are coded up into a computer program. The computer then calculates all the equations using some sort of starting conditions to predict the climate. These models are much more flexible than analogue models and are used to examine past, present and future climates. The difference between the present-day climate simulated by the model and the climate produced by some change in the forcings is the simulated climate change. The development of these models began with the desire to predict the weather, which because of the short time-scales involved, means predicting the condition of the atmosphere.

In 1922 a British meteorologist called Richardson recognized that to predict the weather it would be necessary to solve the physical laws that govern the atmosphere (Richardson, 1922). The basic laws of physics that govern the motion of the atmosphere are called the fundamental equations. They are the conservation of momentum (Newton's second law of motion), the conservation of mass (Continuity equation), the conservation of energy (First law of thermodynamics), the equation of state (Ideal Gas Law) and a conservation equation for moisture. Salby (1992) gives a full mathematical description of these equations and their representation in a rotating frame of reference. What is important to understand is that this group of equations relates the basic variables (pressure, wind velocity, temperature and humidity) which describe the condition of the atmosphere in both space (three dimensions) and time. As the equations are time dependent, given a set of values for the variables at some starting time (the initial conditions), it should be possible to solve the equations for some future time and obtain new values for the variables. Of course the equations would have to be solved in three dimensions. The non-linear equations are extremely complex and approximations have to be derived to solve them. The mathematical techniques for deriving these approximations were available to Richardson. At that time, however, all the calculations would have to have been done by hand. Richardson estimated that it would take a theatre-full of trained meteorologists just to come up with one weather prediction for Britain.

Clearly such a proposal was not possible and weather forecasting had to wait until the arrival of computer technology during the Second World War. Numerical weather prediction models were developed in the post-war years and it is from these that the first climate models evolved. These early models concentrated solely on the atmosphere and are called three dimensional **atmospheric general circulation models** (AGCMs). Until the late 1980s these types of models, although further refined and modified, still dominated climate modelling research. **Oceanic general circulation models** (OGCMs) were developed in the 1960s and since then many other components of the climate system, such as sea-ice and the biosphere, have also been modelled. To simulate the entire climate system this wide variety of models has to be coupled together. In the 1990s, this approach to climate prediction has increased and coupled atmosphere–ocean general circulation models are becoming the primary tools for investigating climate change. However, just like Richardson, climatologists are still limited by their access to computing power. For example a study conducted by the United Kingdom Meteorological Office using such a coupled model took one of the world's fastest supercomputers three months to complete the computation (Hadley Centre, 1995).

Atmospheric general circulation models

Some time will now be spent explaining how atmospheric general circulation models work because they have formed the backbone of many climate studies. Many of the techniques also apply to ocean general circulation models. The atmosphere is an all enveloping body that covers the Earth's surface and the fundamental equations describe in full that continuous nature. AGCMs are given some initial conditions based on current climatological observations. The model is then stepped forward in time with a typical **time step** of 10 minutes. The model solves the fundamental equations to find the values of the variables at this new time. The computation requires the fundamental equations to be approximated. This is done in one of two ways leading to two types of GCM. The first technique is called finite differences and the second technique is called a spectral method. To understand these two techniques a simple analogy will be used before considering some of the mathematics involved. A detailed mathematical description can be found in Hack (1992).

A simple analogy of the finite difference method is to imagine (or try it out) a long string of beads, tightly strung so that there are no gaps. If you lay the beads down into a curve and then move your finger along the beads, from bead to bead in a straight line, the line you would draw with your finger would follow the curve of the string holding the beads together. So even though the beads are individual because they are so small and closely packed together, the line they describe follows the string holding them together. Now imagine that you remove three-quarters of the beads and equally space out the remaining beads along the string. Now when you move your finger from bead to bead in a straight line the line you would draw would probably not follow the curve of the string. It would only approximate the path of the string. The more beads you have the better approximation to the curve of the string. In this analogy of the finite difference method the curve of the string is the mathematical function that we wish to solve and the beads are the solution. We would like to solve it exactly but that would require an infinite number of infinitesimal small beads. Instead we use just a few beads and estimate the function. This analogy transfers simply into mathematics. The string is a curve which can be drawn on graph paper from a mathematical equation as in Figure 6.1.

$$y = f(x) = \cos(x) + \sin(x^2) \tag{6.1}$$

The mathematical equation is said to be a continuous function of x, written f(x). For each value for x, or as in our analogy each bead, you could calculate

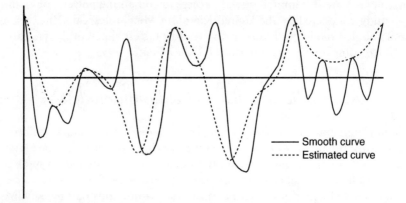

Figure 6.1 The solid line represents the smooth curve which can be approximated by using larger intervals shown by the dashed line

f(x) (the value for the y axis) and draw the line. As the line is continuous, however, to get a good curve you would have to do it for an infinite number of x values. You could draw the line by using just a few values of x, at equal intervals, called discrete points, to calculate the y value. The line drawn, the dashed line in Figure 6.1, is not as smooth or complete as the original but it does approximate it. A disadvantage with this is that to get a good approximation to the continuous function you need a large number of discrete points. It also turns out that to get a workable solution there is a restriction to be met called the **Courant-Freidrichs-Lewy** or **CFL criterion**. In practice it means that as the number of discrete points increases (that is the model resolution increases) the time step has to proportionally decrease. This means the equations have to be solved more often, increasing the computer time needed (Hack, 1992).

The **spectral method** preserves more of the character of the fundamental equations by using mathematical techniques and is consequently harder to describe. Imagine a pile of children's interconnecting building blocks. You are told to build a model of a house with them. At first you are only allowed to use 20 bricks. Your effort might have four walls but it will probably be a very rough approximation to a house. Then you are allowed to use 200, then 2000, then 20 000 bricks and so on. It is not hard to envisage that with larger and larger numbers of building blocks the model you could construct will come closer to looking like a real house. In the analogy, the model house we are trying to construct is the mathematical equation, the more blocks we have, the better the approximation. However, the children's building blocks being used to build the model are an easy-to-use version of real house bricks that are used to construct real buildings. This is the essence of the spectral method; the fundamental equations are transformed by mathematical techniques, into simpler to handle equations, the children's building blocks.

Let's try to put this back into a mathematical context by first explaining the transformation process. The two simplest waveforms are the sine wave and cosine wave as shown in Figure 6.2. Any function which is periodic, that is repeats at constant intervals, can be transformed into a combination of cosine and sine waves. This transformation process is called a Fourier transform. As cosine and sine waves are simple mathematical functions they can be calculated more simply. **The fundamental equations** are periodic and so can be treated in this way. The combination of sine and cosine waves which represent the function form an infinite series. So the entire series, that is all the cosine and sine waves, cannot be calculated. If only 15 are calculated then the series is said to be truncated to wavenumber 15. Thus the spectral method again approximates the fundamental equations, but the solution will be exact up to the truncation limit. If the mathematical function is 'well behaved' then the spectral method is particularly good for approximating the function. However, this is not true of all the equations to be solved. Also if a variable changes very quickly, such as the amount of water vapour, then the spectral method can have difficulties estimating the variable (Hack, 1992).

Using the finite difference method to solve the fundamental equations results in a **Cartesian grid** AGCM. The atmosphere is split up into boxes by a lattice structure, based on the latitude–longitude grid which appears on maps of the Earth. The grid is three-dimensional so extends into the verti- cal as well (*see* Fig. 6.3). The atmosphere in a particular box is represented

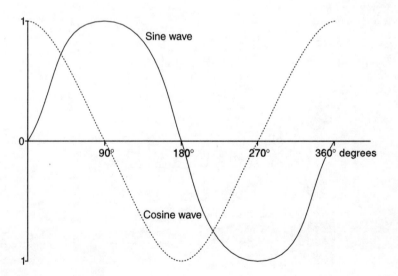

Figures 6.2 The sine and cosine wave function

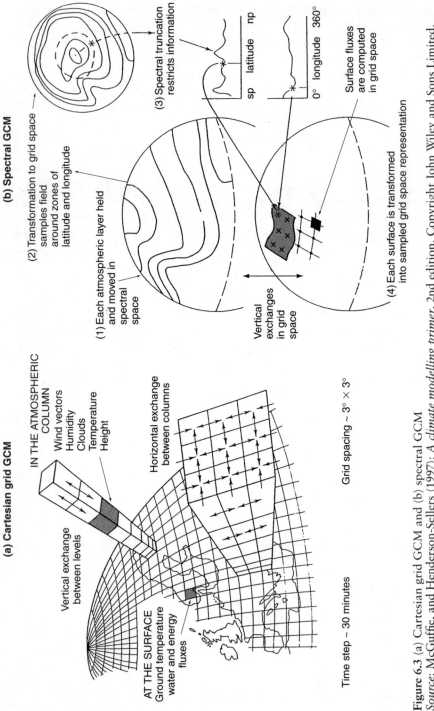

Figure 6.3 (a) Cartesian grid GCM and (b) spectral GCM

Source: McGuffie, and Henderson-Sellers (1997): *A climate modelling primer*, 2nd edition. Copyright John Wiley and Sons Limited. Reproduced with permission.

by the variables at one discrete point at the centre of the box. The fundamental equations are then solved at each grid point. A problem with using the latitude–longitude grid is that near the poles the grid size gets smaller and smaller so to satisfy the CFL criterion the time step must also decrease. Although mathematical techniques can resolve the 'pole problem', it remains to be solved satisfactorily (Hack, 1992). The spectral model, using spectral methods, solves the fundamental equations across a horizontal plane and does not suffer from the pole problem. However, in the vertical plane spectral models most frequently use a finite difference grid to solve the fundamental equations, so time is spent transferring between the two different methods.

Cartesian grid models have a horizontal grid spacing of between 250 to 800 km in both latitude and longitude. The resolution of spectral models is similar. The third dimension, the vertical, in both model types usually has 10 to 20 levels to describe it. However, these are only used to describe the bottom 20 km of the atmosphere. The atmosphere above 20 km altitude has largely been ignored but recent work suggests that for accurate simulations, more of the stratosphere needs to be considered. This will incur increased computational demands. It might be thought that a sensible choice for the vertical co-ordinate would be pressure, however, this leads to some interesting problems as grid points could change height and even appear inside a mountain (McGuffie and Henderson-Sellers, 1997). To avoid these problems a method was devised called the sigma co-ordinate system. In this system the actual pressure at a point is divided by the surface pressure at the point vertically below it. The bottom sigma level follows the contours of the Earth's surface exactly. The resulting surface, however, has to be smoother than reality to avoid errors in the vicinity of steep gradient changes, such as mountains.

Although the fundamental equations may predict the dynamics of the atmosphere, there are other physical processes that also have to be modelled, such as radiation, cloud formation, and surface exchanges. These processes occur on spatial scales far smaller than the grid spacing of the climate model and they are not predicted from first principles. Instead they are parameterized, that is the equations to predict these processes are simplified and formulated to be functions of the variables predicted from the fundamental equations. There are good reasons for **parameterization**. First, it would be impossible computationally to predict everything from first principles. Second, we often do not know in enough detail the underlying principles of the processes and so the equations have to be simplified in some way. Most parameterizations are based on sound empirical evidence or are approximations of more complex functions. An example of the latter is the parameterization of the incoming and outgoing radiation streams. For both Cartesian grid models and spectral models the calculation of these parameterizations is completed on a Cartesian grid. Typically the parameterizations are solved every 3 hours.

Time-dependent climate modelling

Until around the 1990s most computer-based model studies into global warming were 'equilibrium' studies completed on atmospheric general circulation models. Indeed the first IPCC report (Houghton *et al.*, 1990) deals almost exclusively with equilibrium studies. In these studies an AGCM is given as initial conditions a set of observed climatological values. The AGCM is then run until the model settles down. This period is known as the **spin-up time**, and produces a set of variables that should be realistic when compared to present-day climatic averages. These variables present the model's representation of the current climate. In equilibrium studies the amount of carbon dioxide in the atmosphere is then instantaneously doubled. The model is then integrated until a new equilibrium is reached. The difference between the current and predicted model climate represents the model's predictions for how the climate would change given a doubling of carbon dioxide. Such studies are very unrealistic on several accounts. Carbon dioxide is not instantaneously doubling but has been, and will, gradually increase over decades. As long as greenhouse gases continue to rise, the climate will not reach equilibrium but will still be in transition and present a transient climate. Even if carbon dioxide concentrations were to stop rising and hold a constant value it would be decades before a new equilibrium was reached. However, Mitchell *et al.* (1990) point out that there are very good reasons for doing equilibrium studies and these are as follows:

1. They are cheaper because they do not require deep ocean or changes in oceanic circulation and so the models take less computation time.
2. They are easier to compare.
3. In areas of the Earth where the temperature of the ocean changes very slowly equilibrium results can probably be scaled down to give transient results.

The major limitations of equilibrium studies according to Mitchell *et al.* (1990) are

1. They exclude oceanic effects, particularly circulation, or are simplified in some other way (such as using an idealized geography).
2. Different areas of the world will respond at different rates to increasing carbon dioxide concentrations and these may not be realistically simulated with an equilibrium model.

Climate modellers would prefer to do transient or time-dependent modelling with greenhouse gas concentrations rising steadily. Increasingly the climate modellers have moved towards this goal. In order to achieve transient or time-dependent modelling the climate models had to reflect the real world more accurately. This required models to be coupled together. The most important necessity is that the atmospheric general circulation model should be coupled to an oceanic general circulation model. This is so the ocean can be realistically simulated because as greenhouse gas concentrations rise, the heat storage capacity of the oceans will delay any climatic change. At any given time the realized global average temperature will reflect only part of the equilibrium response (Bretherton *et al.*, 1990). The oceans store the rest of the response. Some is stored in the upper ocean (mixed layer) and released in years or decades. The rest is in the deep ocean. It may take centuries for this heating in the deep ocean to influence surface temperatures. Furthermore, the oceans will redistribute the greenhouse warming affecting the equilibrium solutions. Finally fluctuations in the interactions between the atmosphere and ocean will mask longer climatic trends. So for transient studies a coupled ocean–atmosphere model is essential (Bretherton *et al.*, 1990).

Climate modellers have always realized that the only way to truly model the climate system is to couple together the various component models of the climate system. However, as you might have guessed, this is far from simple. To begin with, OGCMs are less well developed than their atmospheric counterparts. Although they began to be developed in 1960s, compared to AGCMs, less attention was initially paid to them (McGuffie and Henderson-Sellers, 1997). Some 70 per cent of the Earth's surface is covered in ocean but there were, and are, few surface observations to provide the data needed for the models. OGCMs have really benefited from the advent of satellite remote sensing and the satellite's ability to provide global coverage. To a large extent OGCMs are based on the same fundamental equations as the atmosphere because both are fluids. However, there are some very important differences. The atmosphere responds much quicker to changes in the energy balance than the ocean. Therefore OGCMs need to consider far longer time-scales than AGCMs. In contrast, the spatial scales of OGCMs need to be much smaller than AGCMs, as important motion occurs at finer spatial scales. Therefore an OGCM is computationally more expensive than an AGCM. The radiation treatment is typically less complex than for the atmosphere. The ocean is also affected far more by the ocean bottom than the atmosphere is by the surface (Haidvogel and Bryan, 1992).

Coupling these two model types together therefore presents many problems. Ideally the two models should share information simultaneously on variables such as energy, momentum and mass exchange. However, given the different spatial and temporal scales this is not possible. An AGCM or OGCM takes a significant amount of computer time to run. Putting the two together is computationally expensive even given the fastest of supercomputers. One way to decrease the computer time needed for a coupled Atmospheric–Oceanic general circulation model (AOGCM) is to use an OGCM with a coarse resolution. However, the important motion which occurs in the ocean at small spatial scales then has to be parameterized. The limitations of coarse resolution OGCMs, which are predominantly used in climate models, are well recognized but are only recently being solved (Gates *et al.*, 1992; Gates *et al.*, 1996). Using a coarse resolution means that the resolution of the OGCM is similar to the AGCM. There remains the problem of the difference in response times between the atmosphere and ocean. To deal with this difference the

AGCM and OGCM do not always share information regularly over the same time period. While the OGCM runs continuously, the AGCM is turned off for periods to save computer time. This mode of operation is called asynchronous coupling. Models that exchange data continuously are said to be fully coupled or synchronous. Some AOGCM model runs contain both synchronous and asynchronous coupling (McGuffie and Henderson-Sellers, 1997).

There are other problems to be overcome when coupling AGCMs and OGCMs which require a more interventionist hand and may well affect the results of coupled AOGCMs (Gates *et al.*, 1996; McGuffie and Henderson-Sellers, 1997). It was noted earlier how a model required 'spin-up' time. In an AOGCM this spin-up time would be equivalent to the response time of the deep ocean – thousands of years. To avoid this climate modellers change the physics of the deep ocean to respond more quickly. The use of this '**distorted physics**', not surprisingly leads to uncharacteristic ocean dynamics and after running with 'distorted physics' the model has to be run for some time without distorted physics to redress the situation. Another well-known problem is that of '**climate drift**'. At local scales differences can occur between the flux values calculated by the AGCM and the OGCM. The errors that result from these differences are large enough to affect the oceanic circulation. The resultant modelled ocean climate essentially drifts away from the observed ocean climate (hence the name). Some modellers accept this but the results show systematic errors. Other modellers apply a **flux adjustment** to stop the drift. Recently several climate models have successfully represented the current climate without the need to resort to flux adjustments (Carson, 1999; Kerr, 1997). Finally there is the '**cold start phenomenon**'. Due to the large heat capacity of the oceans they will respond slowly at first to any warming. In the real world, the build-up of greenhouse gases has occurred since the beginning of the industrial revolution so the slow warming of the ocean has already taken place. In coupled models the forcing is usually applied to present-day (1900) values. For the model to accurately represent the climate it has to account for the warming that has already occurred.

The climate modelling community is not just content to couple ocean and atmosphere GCMs together. Increasingly sophisticated models of other components of the climate system are being developed. It is beyond the scope of this book to discuss them all (see for example, Trenberth, 1992) and the complete features of all the models, as yet, cannot be incorporated into a full **general climate model**. It should be recognized, however, that both the cryosphere and biosphere are being modelled in far more detail than the parameterization schemes that were included in earlier AGCMs. The cryosphere can be split into two sections: sea-ice and land ice. Sea-ice still contains an element of circulation within the modelling but has a two-dimensional nature. The dynamic and thermodynamic properties of sea-ice are strongly coupled and the more complex models consider both of these properties. Land ice, such as glaciers and the polar ice sheets, is still simply modelled in AOGCMs because of the long response times. Dynamic models of ice sheets are yet to be incorporated in AOGCMs. Models of land surface process which take a biophysical approach can be extremely complex but in turn are computationally expensive. Therefore although the modelling of land surface processes within GCMs has improved considerably, simplifications have to be made to the biophysical schemes. Biophysical models typically work on a much smaller grid size than GCMs. A major difficulty for climate modellers is in reconciling these different grid sizes.

Validation or evaluation

Once a climate model has been constructed it needs to be tested. The climate modelling community has widely used the term **validation** to describe this process. However, the IPCC report of 1995 prefers the term **evaluation** (Gates *et al.*, 1996). Most evaluation studies involve comparing the simulated present-day climate to the current observed data. The model will need to simulate accurately the seasonal cycle as well as yearly averages. Furthermore, there is a need to evaluate the model's ability to represent long-term variations in the climate such as the ENSO. Climate modellers will wish to evaluate all aspects of their model. One of the major difficulties in undertaking evaluation studies is the incompleteness of the observational data sets. A full evaluation would require the observational data sets to be obtainable on a wide variety of spatial and temporal scales and these are simply not available. As noted in Chapter 4, surface based observations, while available over a reasonable time span, are limited to hospitable land areas. In contrast, satellite data which provide global coverage have a limited temporal coverage. Several large-scale satellite observational programmes have attempted to redress that situation. However, some variables remain elusive, even to satellite remote sensing. Soil moisture in particular poses problems. Recently large-scale evaluations of general climate models have taken place by intercomparing the models and their components (Gates *et al.*, 1996), such as the Atmospheric Model Intercomparison Project (AMIP) (Gates, 1992) and the Project for Intercomparison of Land surface Parameterization Schemes (PILPS) (Henderson-Sellers *et al.*, 1995).

The evaluation of 11 coupled models is considered in the 1995 IPCC (Gates *et al.*, 1996) report. The report concentrates on evaluating the surface variables, such as temperature and precipitation, which are most pertinent to climate change and to human beings who may have to live with that change. December to February (DJF) and June to August (JJA) results were compared to climatological averages. The 11 coupled models considered by the IPCC did very well at simulating the global scale features of surface air temperature and mean sea level pressure (Gates *et al.*, 1996). This is not to say that there were no disagreements between the models. Part of the process of evaluation is to assess these differences and to isolate the causes. If all the models show the same error then, excluding a problem with the observational data base, this probably reflects a poor understanding and consequently poor representation of one or more of the physical processes in the model. Alternatively, disagreements may stem from certain model practices. For instance, the global average surface air temperature predicted by the 11 models varied by a range of 5°C, however, some of the warmest models did not use a flux adjustment. Those models with a flux adjustment simulated the surface air temperatures more accurately as can be seen in Figure 6.4 (Gates *et al.*, 1996).

The results

This book will not consider equilibrium model results. The reasons why such studies were undertaken in the past have already been outlined. At the time of writing they still form the vast bulk of the studies done and no doubt equilibrium studies will still be run in the future. However, it is clear that time-dependent modelling will increasingly dominate studies. They must be the mode of choice for contemplation of the effects of increased greenhouse gas concentrations. The 1995 IPCC report reflects this change of emphasis (Kattenberg *et al.*, 1996) and the model results presented are almost entirely those run with some sort of coupled model. This section will provide a summary of the current model results. These are as presented by Kattenberg *et al.* (1996) unless otherwise referenced. The IPCC reports contain the widest consideration possible of all the various model results and it is not the purpose, or desire, of this book to repeat these findings. They are used to provide a compact summary and to draw attention to the difficulties faced by those seeking to interpret the model results. The problems will then be discussed in detail in the next section.

In order to run the models some sort of estimate of the emission of greenhouse gases in the future is required. The IPCC use a variety of different greenhouse gas emission scenarios, from low to best guess to high estimates, to obtain model predictions (Table 6.1). The modelling community also recognize that the models themselves reveal a range in how sensitive they are to a change in radiative forcing. Consequently global mean temperature changes are calculated using three different climate sensitivities; a low estimate (1.5°C), best estimate (2.5°C) and a high estimate (4.5°C). Due to the very wide range of scenarios and the time it takes to run a fully coupled OAGCM simpler climate models are used to 'interpolate and extrapolate' between the various scenarios to provide the full range of results (Kattenberg *et al.*, 1996).

Predicted direct effects

For 2100 the predicted global mean temperature increase compared with present day values ranged between 1°C and 4.5°C for the various emission scenarios (Kattenberg, *et al.*, 1996). These are slightly less than that predicted

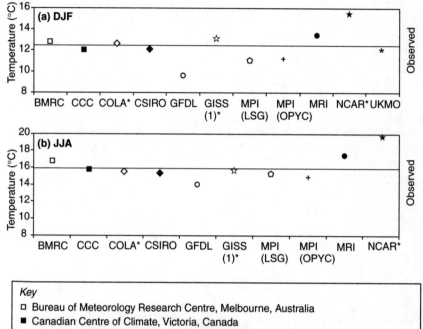

Key
- □ Bureau of Meteorology Research Centre, Melbourne, Australia
- ■ Canadian Centre of Climate, Victoria, Canada
- ◇ Centre for Ocean, Land and Atmosphere, Calverton, USA
- ◆ Commonwealth Scientific and Industrial Research Organization, Melbourne, Australia
- ○ Geophysical Fluid Dynamics Laboratory, Princeton, USA
- ☆ Goddard Institute for Space Studies, New York, USA
- ⬠ Max Planck Institute for Meteorolgy, Hamburg, Germany (LSG)
- + Max Planck Institute for Meteorolgy, Hamburg, Germany (OPYC)
- ● Meteorological Research Institute, Tsukuba, Japan
- ★ National Centre for Atmospheric Research, Boulder, USA
- * United Kingdom Meteorological Office, Bracknell, UK

Figure 6.4 The global average surface air temperature as predicted by coupled general climate models participating in the 1995 IPCC assessment
Source: Gates *et al.* (1996)

Table 6.1 The six scenarios used by the IPCC to estimate emissions of greenhouse gas and other pertinent emissions

Scenario	Population by 2100	Economic growth	Energy supplies	Other	CFCs
IS92a	11.3 B (World Bank 1991)	1990–2025: 2.9% 1990–2100: 2.3%	12 000 EJ Conventional Oil 13 000 EJ Natural Gas Solar costs fall to $0.075/kWh 191 EJ of biofuels available at $70/barrel	Legally enacted and internationally agreed controls on SO_x, NO_x and NMVOC emissions. Efforts to reduce emissions of SO_x, NO_x and CO in developing countries by middle of 21st century	Partial compliance with Montreal Protocol. Technological transfer results in gradual phase out of CFCs in non-signatory countries by 2075.
IS92b	11.3B (World Bank 1991)	1990–2025: 2.9% 1990–2100: 2.3%	Same as 'a'	Same as 'a' plus commitments by many OECD countries to stabilize or reduce CO_2 emissions	Global compliance with scheduled phase out by Montreal Protocol.
IS92c	6.4 B (UN medium-low case)	1990–2025: 2.0% 1990–2100: 1.2%	12 000 EJ Conventional Oil 13 000 EJ Natural Gas Nuclear costs decline by 0.4% annually	Same as 'a'	Same as 'a'
IS92d	6.4 B (UN medium-low case)	1990–2025: 2.7% 1990–2100: 2.0%	Oil and gas same as 'c' Solar costs fall to $0.065/kWh 272 EJ of biofuels available at $50/barrel	Emision controls extended world-wide for CO, SO_x, NO_x and NMVOC. Halt deforestation. Capture and use of emissions from coal mining and gas production and use.	CFC production phase out by 1997 for industrialized countries. Phase out of HCFCs.

Table 6.1 Continued.

Scenario	Population by 2100	Economic growth	Energy supplies	Other	CFCs
IS92e	11.3 B (World Bank 1991)	1990–2025: 3.5% 1990–2100: 3.0%	18 400 EJ conventional oil Gas same as 'a' Phase out nuclear 2075	Emission controls which increase fossil energy costs by 30%	Same as 'd'
IS92f	17.6 B (UN medium-high case)	1990–2025: 2.9% 1990–2100: 2.3%	Oil and gas same as 'e' Solar costs fall to $0.083/kWh Nuclear costs increase to $0.09/kWh	Same as 'a'	Same as 'a'

Source: Leggett *et al.* (1992)

in previous IPCC reports. This is because the emission scenarios now include the effect of aerosols, up to 1990 levels. As explained in Chapter 5, aerosols with an anthropogenic origin, such as sulphate particles caused by industrial activity, are increasing the amount of solar radiation being reflected into space thus cooling the Earth. The incorporation of this effect has allowed the models to reproduce the observed temperature trend (Hasselmann, 1997) and this will be discussed later in this chapter. The overall effect is that global warming is ameliorated, slightly. Of course, just as with greenhouse gases, it is unclear how the amount of aerosols in the atmosphere will vary in the future. Consequently, temperature change has also been estimated with a variety of emission scenarios for aerosols in the future from 1990 levels. These temperature changes are lower still, being between 1°C and 3.5°C by 2100. However, as stressed by the IPCC summary this is still a substantial temperature change (Kattenberg *et al.*, 1996). Figure 6.5 shows typical climate model temperature results for the IS92a scenario from the UK Hadley Centre, Meteorological Office (Mitchell *et al.*, 1995). The model is a coupled model and Figure 6.5a shows the surface air temperature changes predicted from increasing carbon dioxide. Figure 6.5b shows the effect on the temperature changes of including sulphate aerosols. Kattenberg *et al.* (1996) also considered the impact on the global temperature if greenhouse gases were stabilized under certain emission scenarios. In all cases the global temperature continued to rise during the twenty-first century.

Land areas are predicted to warm more than the oceans. The maximum warming occurs at high northern latitudes and is greatest during late autumn and winter. The inclusion of aerosols lessens both of these effects in the northern latitudes. There is no seasonal variation in temperatures at low latitudes or in the southern Antarctic ocean. The difference in temperature between night and day over land areas is decreased in the majority of seasons and in most regions.

There is an increase in global mean precipitation which is a result of the increased temperature. The inclusion of aerosols consequently decreases the amount of precipitation. Precipitation increase is greatest at high latitudes during winter and this also affects the mid-latitudes. Other effects are less clear with models showing varying results such as changes in the amount of tropical rainfall. Although all the models show increased soil moisture in high northern latitudes in winter and a decrease in summer there is less confidence in these results compared to temperature. The vigour of the northern North Atlantic oceanic circulation is decreased due to the increased precipitation at high latitudes decreasing salinity and inhibiting the sinking of water at these latitudes. This reduces the amount of warming in the North Atlantic. Coupled models have revealed that threshold values of atmospheric carbon dioxide occur at which the oceanic conveyor belt stops. Schmittner and Stocker (1999) found results with a simplified coupled model suggest that the rate of increase of carbon dioxide may also be an important factor in the slowing of the conveyor belt.

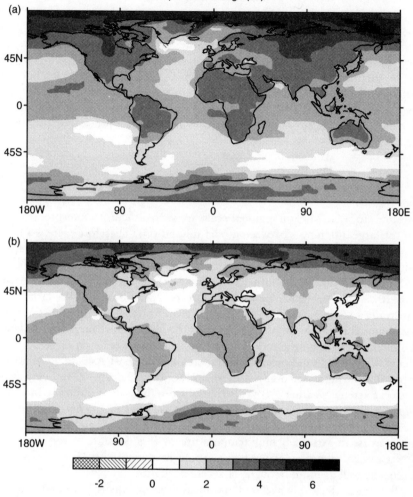

Figure 6.5 Coupled general climate model results from the Hadley Centre Model, Meteorological Office, UK (a) Surface air temperature changes due to increasing greenhouse gases only (IS92a emission scenario) (b) as for (a) only with the inclusion of sulphate aerosols
Source: Mitchell *et al.* (1995). Reprinted with permission from *Nature.* Mitchell, J.F.B., Johns, T.C., Gregory, J.M. and Tett, S.F.B.: Climate response to increasing levels of greenhouse gases and sulphate aerosols. *Nature* **376**, 501–504, 1995: Macmillan Magazines Ltd. © Crown copyright. Reproduced by permission of the controller of HMSO.

Predicted indirect effects

Perhaps the most widely considered indirect effect is that of sea level rise. The amount of sea level rise predicted depends on the emission scenario used and the model uncertainties. The largest range of uncertainties was 13 to 94 cm rise with a best estimate guess of 38 to 55 cm by 2100. This was mostly due to thermal expansion of the oceans and then from the melting of glaciers and ice caps. Over this time period contributions from the ice sheets are thought to be small but have associated with them the greatest uncertainty. Gregory and Oerlemans (1998) calculated a 132 mm sea level rise due to glacier melt between 1990 and 2100 and another 70 mm from the melting of the Greenland Ice Sheet. This is 20 per cent higher than previous estimates of glacier melt.

Regional scenarios

Results from GCMs at a regional scale are very difficult to interpret because of the large differences between model simulations. Therefore there is a lot of uncertainty associated with any of the predictions. The 1990 IPCC report identified five regions for analysis: these were Central North America (35–50°N, 85–105°W), South East Asia (3–30°N, 70–105 °E), Sahel, Africa (10–20°N, 20°W–40°E), Southern Europe (35–50°N, 10°W–45°E) and Australia (12–45°S, 110–155°E). Temperature differences between the transient model results were often as large as the temperature increase predicted. Differences between the models over precipitation were even larger, though generally it was found precipitation increased over East Asia and wintertime precipitation increased over Northern and Southern Europe (Kattenberg *et al.*, 1996). Extreme weather events also suffer from the lack of ability to model at small scales. This is because many extreme weather events occur on spatial scales much smaller than that of a GCM. However, as might be expected, an increase in extremely high temperatures with a decrease in very low temperatures in winter were predicted. Some models predicted more flooding events as well as more frequent or severe droughts.

For policy decisions national and local governments require national and local predictions. Yet global climate models are at their weakest when it comes to predicting climate changes on these sorts of spatial scales. Therefore there has been increasing emphasis on producing models which operate at smaller spatial scales. The 1995 IPCC report (Houghton *et al.*, 1996) and the recent literature reflects this determination to produce results at a higher resolution. Regional scale modelling uses GCM results to predict changes on these smaller scales. The difficulties of **downscaling** will be discussed in the next section. Studies with regional scale models seem to confirm some of the GCM findings such as an intensification of the hydrological cycle. Knutson and Tuleya (1999) found that a hurricane prediction model predicted greater hurricane intensity with increased carbon dioxide levels. Other studies over Europe find an increase in the frequency of high precipitation events (Arnell, 1999; Frei *et al.*, 1998; Jones *et al.*, 1997).

Regional scale modelling

The problem that modellers face in this research area is how to downscale the results of general climate models to the regional scale. There are two ways to achieve this downscaling: one is an empirical approach using a variety of mainly statistical methods and the other a process approach whereby a high resolution computer model is nested inside a general climate model. Both methods use the results from general climate models to drive the higher resolution model.

Empirical downscaling

There are several ways in which empirical models can be generated to provide local scale climate predictions. Earlier it was explained how general climate models use the fundamental equations to predict pressure, wind velocity, temperature and humidity. These variables are sometimes referred to as the large-scale variables. The first stage in the empirical downscaling technique is to develop relationships between local observed climate variables, such as surface air temperature and precipitation and the large-scale variables. Having established the relationship, the second stage is to use the general climate model output for doubled carbon dioxide as the predictor variables in the empirical model to estimate the local perturbed variables. To illustrate this consider a general climate model with a resolution of $2.5°$ by $2.5°$, running to simulate the present-day climate. At the surface the model will produce just one value for the temperature for the whole grid box. Within that grid box, however, there might be six surface stations with observers each taking a temperature reading. Due to the local variations in climate those surface observed readings will often vary from one another. Empirical downscaling requires a relationship to be developed between the one model temperature result and the six surface observed values. This is achieved by using several years of data to, at the simplest, perform a regression analysis between the model and observed data. Model temperature data from doubled carbon dioxide runs can then be used as input into the regression equations derived to predict the local temperature at each of the surface stations. Regression methods are amongst the simplest techniques to be used. Neural networks can also be used and build up a model in much the same way as regression analysis (Wilby and Wigley, 1997). There are also more complex approaches in which observed surface data are related to weather patterns by either sophisticated statistical or neural network methods (Wilby and Wigley, 1997).

Regional modelling

Regional climate models (RCMs) are computer models which predict the regional climate from first principals. Sometimes called **limited area models** (LAMs) they have a horizontal grid spacing of between 20 to 50 km and a vertical resolution between 100 and 1000 m which is much smaller than most general climate models (Figure 6.6). The output from the general climate model drives the smaller regional model. Imagine the box grid-like structure of the Cartesian general atmospheric models (Figure 6.3). The RCM is placed within one of the boxes. To the sides and both above and below the RCM are the large-scale boxes of the GCM, producing large-scale variable results. These results act as the initial conditions for the RCM. The GCM also provides the driving lateral meteorological boundary conditions (Kattenberg *et al.*, 1996). That is, the RCM is constrained at the boundaries with the GCM to those conditions set by the GCM. These models are called one-way nested models because although the GCM results drive the RCM there is no feedback from the RCM into the GCM. By nesting RCMs inside GCMs a much smaller spatial scale can be achieved and this allows the investigation of sub-grid scale GCM forcings to be considered (Kattenberg *et al.*, 1996). RCMs,

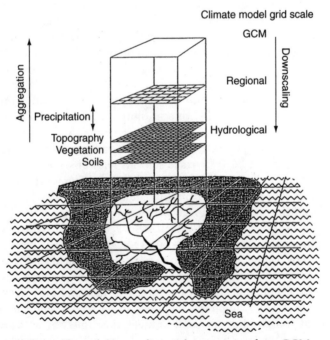

Figure 6.6 The problem of downscaling and aggregation from GCMs to hydrological models
Source: Wilby and Wigley (1997)

however, require as much computer resources as a GCM so are expensive to run. They are also very dependent on the quality of the input from the GCM (Wilby and Wigley, 1997).

The problems and the criticisms

A new ice age

The simplest computer-based climate models are energy balance models (EBMs). These models divide the Earth into latitude zones. As latitude is the only dimension used these are one-dimensional models. For each latitude zone the model computes the incoming and outgoing energy. The equations that govern the model are all written in terms of the variable to be predicted, usually surface air temperature. This type of energy balance model came to prominence in the late 1960s when they were used to assess the sensitivity of the climate to a change in the amount of incoming solar radiation. The models predicted that a relatively small decrease in the amount of radiation reaching the Earth could lead to total glaciation. At the same time as these results were published, northern hemisphere temperatures were on a downward trend. This led to speculation that the Earth could be heading for a new 'ice age'. In the 1970s it was global cooling that dominated the text books and grabbed media attention; nor was it all hype. As Grove (1988) points out, the northern hemisphere cooling and glacial advance at this time had distinct financial implications for several northern European countries.

Arrhenious in 1908 and Callendar in 1938 had proposed that increasing carbon dioxide might save the world from a new ice age and these words were echoed again in the 1970s (Fleming, 1998; Mitchell, 1972). The extreme sensitivity of energy balance models, however, turned out to be due to the simple way the model represented the climate and not a true picture of the climate response. Energy balance models have increased in sophistication and are still in use today, mostly to study climates in the geological past. The legacy of these early modelling results remain, however, to haunt the present. It is also true that we are on the downturn of the next glacial cycle and are heading towards a new glacial maximum. Past inter-glacials have lasted between 10 000 and 14 000 years and this one is already 10 000 years old. However, to cite this as a rationale for not worrying about global warming is to forget about time-scales. Global warming will occur over the next few decades to hundred of years. Furthermore, our inadequate understanding of the interplay between radiative forcings and climatic responses should caution against relying on any such simplistic balancing act between global warming and a supposed natural cooling.

Future emissions

Climate models require accurate assessments of the amount of greenhouse gas emissions in the future if they are to accurately predict the resultant climatic change. A major problem is that it is almost impossible to correctly predict the amount of anthropogenic greenhouse gases that will accumulate in the atmosphere over this period. This is because emission scenarios depend not only on our sometimes incomplete knowledge of natural sources and sinks of greenhouse gases, but also on socio-economic predictions. Emission scenarios contain estimates of the population growth and try to take into account economic growth and use of renewable technology. The 1990 IPCC report used several scenarios which were then updated for the 1992 interim report (Houghton *et al.*, 1992) and these remain in use for the 1995 report (Houghton *et al.*, 1996). The revision in 1992 took into account changes in the political structure of eastern Europe and the Montreal Protocol developments. Table 6.1 lists the six scenarios formed based on various assumptions. The scenario IS92a most closely corresponds to the 'Business as Usual' scenario in the 1990 IPCC report. This essentially takes into account 1995 legislation affecting air pollution but does not include controls on carbon dioxide. The 1992 IPCC report is at pains to point out that none of the scenarios are really predictions of the future and that the farther we look ahead the more likely they are to be wrong. The emission scenarios merely provide input into the models to help policy-makers decide on strategies and to improve our understanding (Leggett *et al.*, 1992). Gray (1998) claims that all the IPCC scenarios exaggerate at least one of the trends they are trying to predict. That this in turn leads to excessive greenhouse gas concentrations being predicted. Furthermore, the inaccuracies are so great that they are useless in terms of policy predictions. Hulme (1999), however, rejects Gray's claim presenting climate model predictions for the 1990s that are in line with the observed temperature trend.

The models

In discussing the development of general circulation models numerous difficulties, some of which still afflict model results, have been mentioned. Model's treatment of the cold start phenomenon and climate drift will affect its results. Anyone studying the model output should be aware of these problems and the solutions applied in the model in order to correctly interpret the model results. Being aware of your data limitations is an important starting point, but there are those who would question whether we should have any confidence in model results at all. Aspects of modelling have improved with coupled OAGCMs allowing transient climate studies to be completed. However, some of the most severe criticisms of the predictions regarding global warming are about the details of the models. The modellers themselves recognize the limitations of even the most current and advanced models. A major difficulty for those of us outside the modelling community is recognizing genuine criticisms of the models and placing some uncertainty estimate upon them.

Most of the models appear to predict the current climate with some degree of accuracy. However, it might be argued that they are tuned to that result. Early model simulations relied on prescribed sea surface temperatures and the most accurate coupled models rely on a flux adjustment to correct climate drift and return back to the norm. Given that the oceans cover 70 per cent of the Earth's surface, this is a large area to adjust. The development of coupled climate models without the flux adjustment should lead to significantly improved results (Carson, 1999). It is also difficult to know just how each model varies in its computer coding. There are a limited number of ways in which you can realistically model the Earth's climate. There are two standard ways of representing convection in an atmospheric general circulation model, the Kuo scheme and Betts-Miller Scheme. So although there may be 30 leading computer modelling centres they only use one of two schemes for convection, which leads to the question, just how different are all these models? There are, however, hundreds of different parameterizations included in every model and each will be coded up in a unique way. As a result, it is extremely difficult to take a component part of one model and run it in another host model without having to change other parts of the host model. Therefore there is an inability to replicate results which is usually a pre requisite of scientific experiments.

Many critics of the models centre their objections on the inability of the models to reproduce the observed warming trend (Pearce, 1997). The observed warming trend has been slower than most models predicted and this discrepancy is a main point of contention between the sceptics and the modellers. The modellers assert that the difference is due to cooling effects that are masking global warming. The problem is to identify these cooling effects and include them in the model. Eventually the greenhouse effect will dominate and global warming will occur. The sceptics argue that the difference between the model results and the observations is due to an overestimation of the positive feedbacks in the climate system and hence the sensitivity of the climate to increased greenhouse gases (Pearce, 1997).

The modellers have had some success in identifying cooling effects in the atmosphere. One of the main culprits is thought to be aerosols from industrial activity. The Hadley Centre at the UK Meteorological Office modelled climate change from 1860 to 2050 using their Unified Model. The experiment consisted of using a coupled atmosphere–ocean model in a transient experiment which included not only increasing greenhouse gases but also the sulphate aerosol. The main finding of this experiment was that the model temperature trend was now in line with the observed temperature trend (Hadley Centre, 1995). Since then other modellers have found similar results (Hasselmann, 1997). Unfortunately this cooling effect is unlikely to save the

Earth from global warming. The aerosol haze is patchy, occurring predominantly in areas of high industrial activity, such as the northern hemisphere, so tends to have a localized effect on the radiation balance. Also there are desperate efforts to decrease aerosol pollution particularly sulphur production from power stations which cause acid rain. In addition, aerosols are much shorter-lived in the troposphere than carbon dioxide and respond to changes in emissions in a few weeks compared to 10 to 100 years for carbon dioxide. Therefore reduction in aerosol emissions will soon lead to a reduction in the sulphate haze and a quick return to more natural levels (Kerr, 1992).

These two aspects of sulphate aerosols, their localized nature and short time-scales, have led some to criticize even this apparent success of the models. It has been argued that as most of the industrial pollution emanates from the northern hemisphere then the cooling should be limited to the north (Pearce, 1997). The southern hemisphere should warm faster than the north and this occurred until 1987. However, since then, Patrick Michaels (Pearce, 1997) claims warming in the southern hemisphere has ceased while in the northern hemisphere it has leapt ahead. Consequently model results no longer reflect the observations. Not surprisingly this has been heavily contested with counter-claims that this is accounted for by model results (Pearce, 1997). It has been suggested that these modelling results which include sulphate aerosols over-play the effect of the aerosols and that the cooling effect of aerosols is in fact much smaller than those used in the models. This leads some authors to suspect that there are other moderating effects that have yet to be incorporated in the models (Pearce, 1995). Hansen *et al.* (1997) argue that the aerosol effect on clouds is far larger than has been assumed. Therefore the indirect cooling effect of aerosols through changing cloud properties is much greater and leads to a large negative forcing. Cane *et al.* (1997) suggest that a trend in sea surface temperatures observed in the data and predicted theoretically but not the models may account for some of the discrepancy between observed and modelled temperature trends in the twentieth century.

Until recently many climate model results predicted that the upper troposphere would warm faster than the surface. This is exactly opposite to the results received from radiosondes and microwave satellite measurements which have found no warming in the upper atmosphere (Bengtsson *et al.*, 1999). However, most atmospheric climate models ignored the atmosphere above the troposphere, now the stratosphere is being included in order to assess the impact of stratospheric ozone depletion on the vertical temperature profile of the atmosphere. With a reduction of stratospheric ozone the stratosphere will cool, as ultraviolet radiation will no longer be absorbed there. By the Stefan-Boltzman law the stratosphere will then radiate less heat upwards and downwards into the troposphere. Therefore the upper troposphere will cool. By including this effect modellers claim that model results are in line with observations (Bengtsson *et al.*, 1999; Pearce, 1997). Again sceptics decry this continual altering of the models to fit observations as unscientific (Pearce, 1997).

Arguments that surround the sensitivity of the climate to increasing greenhouse gases centre on feedbacks and in particular the role of water vapour (Lindzen, 1997). Models predict an aggressive positive feedback associated with water vapour. Global warming is predicted to result in more evaporation from the oceans, releasing more water vapour into the atmosphere. As water vapour is a greenhouse gas, this will amplify the affect of increased anthropogenic greenhouse gases. Sceptics take issue with this view and their arguments are based on the importance of the vertical profile of the water vapour in the troposphere. It is in the free troposphere where water vapour has to increase for a significant positive feedback to take place. Sceptics claim that the observational evidence suggests this will not occur and most water vapour will end up near the surface where the positive feedback effect will be small. Lindzen (1994) argues that in a warmer world air parcels in clouds would be more buoyant so they could rise to higher altitudes where it is colder. As the air is colder it can hold less water vapour. Thus the amount of water vapour in the free atmosphere would decrease. Given that water vapour is the most potent greenhouse gas, Lindzen (1994) concludes that this would be a negative feedback reducing global warming. Furthermore, sceptics argue that warmer clouds will be more efficient at producing rain leading to less re-evaporation from the clouds and therefore less water vapour in the free troposphere (Pearce, 1997). The arguments for a positive feedback and negative feedback from clouds are shown in Figure 6.7. Much of the discussion is centred on what happens to cloud amounts in the upper troposphere in the tropics which is known to be extremely dry (Spencer and Braswell, 1997). Modelling studies suggest that with global warming the troposphere becomes higher with more water vapour leading to increased amounts of high cloud which will act as a positive feedback to global warming (Dickinson *et al.*, 1996). It is generally acknowledged that the treatment of water vapour and clouds in climate models are areas of great uncertainty. This is because clouds occur at scales which are smaller that the climate model grid. Therefore they are frequently included as a parameterization scheme.

The cloud problem

As explained in Chapter 2, clouds are very important in that they modify the radiation balance of the Earth in two ways. Due to their high albedo they reflect incoming radiation and thus act to cool the Earth. Being composed of water, however, they also absorb outgoing long-wave radiation acting to warm the Earth. Whether a cloud warms or cools the surface depends on many factors; its geographical position, height, the size of the water droplets. This radiation balancing act of clouds means that they have a strong influence on the radiation budget of the Earth. Some studies indicate that an increase in cloud cover of just a few percentage could offset the effects of doubling carbon dioxide in the atmosphere (Coley and Jonas, 1999). What will

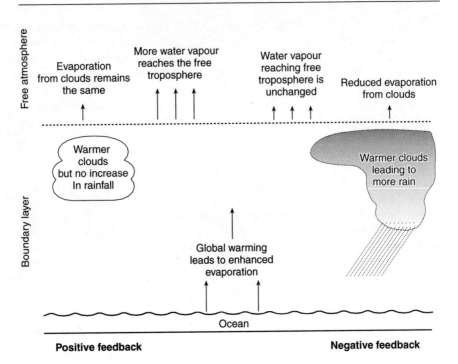

Figure 6.7 Competing claims as to whether global warming will lead to a water vapour feedback which is positive leading to greater warming or negative leading to cooling as the sceptics claim
Source: After Pearce (1997)

happen in a warmer world, whether cloud cover or the cloud radiative properties will change, is a further source of debate (Lindzen, 1994). As indicated in the previous section, increased cloud cover seems likely but the exact properties of the clouds will strongly determine their influence on the Earth's radiation regime. The limited knowledge we have about cloud properties hampers the modelling of clouds.

Clouds are poorly defined in many ways. Surface observations of clouds rely on a subjective assessment of cloud amount and height and are a measure of sky cover. They fail to provide any detailed radiative properties of the clouds. Satellite observations of clouds are based on the interpretation of radiance measurements and are a ground cover measurement. So surface and satellite observations of clouds are two different things which are very difficult to compare. The layered nature of clouds further complicates the picture and neither surface nor satellite measurements, so far, have been able to reveal the full detail of cloud properties. Modelled clouds are different again and their main impact within climate models until recently has been to modify the incoming and outgoing radiation streams. More recently models have been predicting cloud liquid water content directly from the fundamental equations rather than as parameterizations and this should lead to an improvement in the simulation of the Earth's radiation budget (Ghan *et al.*, 1997; Kattenberg *et al.*, 1996; Senior and Mitchell, 1993).

Regional difficulties

General circulation model resolution is of the order of 200km, which means that 10 grid squares or less cover the whole of Britain. This does not result in very good local detail. Therefore, confidence in the regional predictions of climate models is significantly lower than at the global scale. Thus for areas the size of North America the model results differ significantly. This in itself creates a problem for planning. How does a country the size of Britain interpret model results? This problem has been at the centre of many criticisms of global warming predictions. The model results are not good enough at local scales for firm policy decisions to be made. Therefore to answer this criticism and to understand more about the regional changes that may occur with global warming, considerable effort is being expended in downscaling GCM results. All the various methods used for downscaling GCM output have their limitations. Wilby and Wigley (1997) list three major challenges that have to be overcome by downscaling techniques. First, classification techniques have to be developed that can be applied globally rather than just to one local area. Second, the method has to capture the variability of the climate at all spatial and temporal scales. Currently downscaling techniques fail to do this. Third, and most importantly, the relationships between the large scale and the local scale have to be shown to be stable and not vary from year to year. Downscaling techniques are still in comparative infancy and need to be developed further. Comparisons of the various techniques would seem to conclude that, at present, the results obtained from empirical methods are as good as those from RCMs especially given the additional expense of RCMs (Kidson and Thompson, 1998).

The oceans

When discussing time-dependent modelling earlier in this chapter a lot of attention was paid to the difficulties of coupling atmospheric and oceanic models. It was also noted that oceanic modelling still lagged behind atmospheric modelling. In previous chapters the importance of the oceanic conveyor belt to the climate system and possible links to quaternary glacial cycles have also been stressed. A breakdown of the oceanic conveyor belt in the North Atlantic due to an influx of fresh water is thought responsible for pushing Europe back into the last ice age. If global warming was to cause melting of the polar ice caps, could this also lead to a shut down of the oceanic conveyor? Instead of warming, could Europe experience a sudden cooling (Bunyard, 1999)? The models have shown that warming could stop the conveyor belt (Schmittner and Stocker, 1999). However, as yet the models are not sophisticated enough to detail the level of warming required. This is due to the coarse resolution used, poor treatment of salt and heat diffusion and the flux adjustment needed to address climatic drift (Rahmstorf, 1997). There is a lot of evidence to suggest that the oceanic conveyor belt is unstable. This may, however, only be a feature of the oceanic circulation when there are large amounts of land ice. To try to unravel this, the last interglacial during the Eemian (113 000 to 125 000 years ago) is being studied. Ice cores from the GRIP and GISP2 projects initially provided conflicting results, with the former suggesting large climatic changes. It has now been ascertained that this was due to disruption of the ice and the GISP2 core, suggesting a quiet climate. Furthermore, a study of oceanic sediments also suggest that the Eemian interglacial was a period of climatic stability (Adkins *et al.*, 1997).

Summary

Depending on your stance on global warming climate models either provide the best picture of climate change yet available or are a mirage of scientific fact, providing only the illusion of certainty rather than anything concrete. In 1896 Arrhenious published his now famous paper on the influence of carbonic acid (carbon dioxide) in the atmosphere. He used a very simple energy balance model to calculate the change in temperature at each latitude band for a given change in carbon dioxide. Arrhenious was really more concerned with what caused the quaternary glacial cycles and initially dismissed anthropogenic carbon dioxide emissions as irrelevant, but the conclusions of his paper are none the less interesting. The temperature change is smallest at the equator and less in the southern hemisphere compared with the northern hemisphere. The effect is greater in winter than summer and also greater for land than for the ocean. Also, the temperature difference between day and night will be reduced. If you look at the general conclusions drawn from the IPCC in 1995, a 100 years on from the Arrhenious paper, it would appear that not much has changed. However, this fails to recognize the scientific knowledge and wealth of detail that have been gained in those 100 years. Climate models have changed immeasurably and the advancements made have considerably improved our ability to predict the climate of the future (Carson, 1999). This progress, however, would still appear to be outstripped by the requirements of policy-makers and the changing perception of global warming from an interesting scientific problem to a political challenge. Computer-based models are a tool to aid our understanding of the climate, but that is not what policy-makers need. They require information on the future climate in order to predict the impact on economic and social development. Governments require that sort of information to implement change where needed. The final chapter considers the implications of the need for scientific certainty in environmental studies and the relationship between science, environment and politics. The next chapter looks at ways in which the social dimensions of global warming are assessed.

References

Adkins, J.F., Boyle, E.A., Keigwin, L. and Cortijo, E. 1997: Variability of the North Atlantic thermohaline circulation during the last interglacial period. *Nature* 390, 154–6.

Arnell, N. 1999: The effects of climate change on hydrological regimes in Europe: a continental perspective. *Global Environmental Change* 9, 5–23.

Bengtsson, L., Roeckner, E. and Stendel, M. 1999: Why is global warming proceeding much slower than expected? *Journal of Geophysical Research-Atmospheres* 104, 3865–76.

Bretherton, F.P., Bryan, K. and Woods, J.D. 1990: Time-dependent greenhouse-gas-induced climate change. In Houghton, J.T., Jenkins, G.J. and Ephraums, J.J. (eds), *Climate change: the IPCC scientific assessment.* Cambridge: Cambridge University Press.

Bunyard, P. 1999: How global warming could cause northern Europe to freeze. *The Ecologist* 29, 79–80.

Cane, M.A., Clement, A.C., Kaplan, A., Kushnir, Y., Pozdnyakov, D., Seager, R., Zebiak, S.E. and Murtugudde, R. 1997: Twentieth-century sea surface temperature trends. *Science* 275, 957–60.

Carson, D.J. 1999: Climate modelling: achievements and prospects. *Quarterly Journal of the Royal Meteorological Society* 125, 1–27.

Coley, P.F. and Jonas, P.R. 1999: Clouds and the Earth's radiation budget. *Weather* 54, 66–70.

Dickinson R.E., Meleshko, V., Randall, D., Sarachik, E., Silva-Dias, P. and Slingo, A. 1996: Climate processes. In Houghton, J.T., Meira Filho, L.G., Callander, B.A., Harris, N., Kattenberg, A. and Maskell, K. (eds), *Climate change 1995: the science of climate change.* Cambridge: Cambridge University Press.

Fleming, J.R. 1998: Arrhenious and current climate concerns: continuity or 100-year gap? *Eos* 79, 405, 409, 410.

Frei, C., Schar, C., Luthi, D. and Davies, H.C. 1998: Heavy precipitation processes in a warmer climate. *Geophysical Research Letters* 25, 1431–4.

Gates, W.L. 1992: AMIP: the Atmospheric Model Intercomparison Project. *Bulletin of the American Meteorological Society* 73, 1962–70.

Gates, W.L., Henderson-Sellers, A., Boer, G.J., Folland, C.K., Kioth, A., McAvaney, B.J., Semazzi, F., Smith, N., Weaver, A.J. and Zeng, Q.-C. 1996: Climate models – evaluation. In Houghton, J.T., Meira Filho, L.G., Callander, B.A., Harris, N., Kattenberg, A. and Maskell, K. (eds), *Climate change 1995: the science of climate change.* Cambridge: Cambridge University Press.

Gates, W.L., Mitchell, J.F.B., Boer, G.J., Cusbach, U. and Meleshko, V.P. 1992:

Climate modelling, climate prediction and model validation. In Houghton, J.T., Callendar, B.A. and Varney, S.K. (eds), *Climate change 1992: the supplementary report to the IPCC scientific assessment*. Cambridge: Cambridge University Press.

Ghan, S.J., Lueng, L.R. and Hu, Q. 1997: Application of cloud microphysics to NCAR community model. *Journal of Geophysical Research-Atmospheres* **102**, 16507–27.

Gray, V. 1998: The IPCC future projections: are they plausible? *Climate Research* **10**, 155–62.

Gregory, J.M. and Oerlemans, J. 1998: Simulated future sea-level rise due to glacier melt based on regionally and seasonally resolved temperature changes. *Nature* **391**, 474–6.

Grove, J.M. 1988: *The Little Ice Age*. London: Methuen.

Hack, J.J. 1992: Climate system simulation: numerical and computational concepts. In Trenberth, K.E. (ed.), *Climate system modeling*. Cambridge: Cambridge University Press.

Hadley Centre 1995: *Modelling climate change 1860-2050*. London: Meteorological Office.

Haidvogel, D.B. and Bryan, F.O. 1992: Ocean general circulation modelling. In Trenberth, K.E. (ed.), *Climate system modeling*. Cambridge: Cambridge University Press

Hansen, J., Sato, M., Lacis, A. and Ruedy, R. 1997: The missing climate forcing. *Philosophical Transactions of the Royal Society of London, Series B-Biological Sciences* **352**, 231–40.

Hasselmann, K. 1997: Are we seeing global warming? *Science* **276**, 914–15.

Henderson-Sellers, A., Pitman, A.J., Love, P.K., Irannejad, P. and Chen, T.H. 1995: The Project for Intercomparison of Land surface Parameterisation Schemes (PILPS): Phases 2 and 3. *Bulletin of the American Meteorological Society* **76**, 489–503.

Houghton, J.T., Callander, B.A. and Varney, S.K. (eds) 1992: *Climate change 1992: a supplementary report to the IPCC scientific assessment*. Cambridge: Cambridge University Press

Houghton, J.T., Jenkins, G.J. and Ephraums, J.J. (eds) 1990: *Climate change: the IPCC scientific assessment*. Cambridge. Cambridge University Press.

Houghton, J.T., Meira Filho, L.G., Callander, B.A., Harris, N., Kattenberg, A. and Maskell, K. (eds) 1996: *Climate change 1995: the science of climate change*. Cambridge University Press. Cambridge.

Hulme, M. 1999: Global warming. *Progress in Physical Geography* **23**, 283–91.

Jones, R.G., Murphy, J.M., Noguer, M. and Keen, A.B. 1997: Simulation of climate change over Europe using a nested regional-climate model 2: comparison of driving and regional model responses to a doubling of carbon dioxide. *Quarterly Journal of the Royal Meteorological Society* **123**, 265–92.

Kattenberg, A., Giorgi, F., Grassl, H., Meehl, G.A., Mitchell, J.F.B., Stouffer, R.J., Tokioka, T., Weaver, A.J. and Wigley, T.M.L. 1996: Climate models –

projections of future climates. In Houghton, J.T., Meira Filho, L.G., Callander, B.A., Harris, N., Kattenberg, A. and Maskell, K. (eds), *Climate change 1995: the science of climate change*. Cambridge: Cambridge University Press.

Kerr, R.A. 1992: Pollutant haze cools the greenhouse. *Science* 255, 682–3.

Kerr, R.A. 1997: Model gets it right – without fudge factors. *Science* 276, 1041.

Kidson, J.W. and Thompson, C.S. 1998: A comparison of statistical and model-based down scaling techniques for estimating local climate variations. *Journal of Climate* 11, 735–53.

Knutson, T.R. and Tuleya, R.E. 1999: Increased hurricane intensities with CO_2 induced warming as simulated using the GFDL hurricane prediction system. *Climate Dynamics* 15, 503–19.

Leggett, J., Pepper, W.J. and Stewart, R.J. 1992: Emission scenarios for IPCC: an update. In Houghton, J.T., Callandar, B.A. and Varney, S.K. (eds), *Climate change 1992: the supplementary report to the IPCC scientific assessment*. Cambridge: Cambridge University Press.

Lindzen, R.S. 1994: On the scientific basis for global warming scenarios. *Environmental Pollution* 83, 125–34.

Lindzen, R.S. 1997: Can increasing carbon dioxide cause climate change? *Proceedings of the National Academy of Sciences of the United States of America* 94, 8335–42.

McGuffie, K. and Henderson-Sellers, A. 1997: *A climate modelling primer*. 2nd edition. Chichester: John Wiley and Sons.

Mitchell, J.F.B., Johns, T.C., Gregory, J.M. and Tett, S.F.B. 1995: Climate response to increasing levels of greenhouse gases and sulphate aerosols. *Nature* 376, 501–4.

Mitchell, J.F.B., Manabe, S., Meleshko, V. and. Tokioka, T. 1990: Equilibrium climate change – and its implications for the future. In Houghton, J.T., Jenkins, G.J. and Ephraums, J.J. (eds), *Climate change: the IPCC scientific assessment*. Cambridge: Cambridge University Press.

Mitchell Jr., M.J. 1972: The natural breakdown of the present interglacial and its possible intervention by human activities. *Quaternary Research* 2, 436–45.

Pearce, F. 1995: Experts blow cold on climate claims. *New Scientist* 148 (no. 2003), 5.

Peace, F. 1997: Greenhouse wars. *New Scientist* 155 (no. 2091), 38–43.

Rahmstorf, S. 1997: Ice-cold in Paris. *New Scientist* 153 (no. 2068), 26–30.

Richardson, L.F. 1922: *Weather prediction by numerical processes*. Cambridge: Cambridge University Press, (Reprinted by Dover, 1965).

Salby, M.L. 1992: The atmosphere. In Trenberth, K.E. (ed.), *Climate system modeling*. Cambridge: Cambridge University Press.

Schmittner A. and Stocker, T.F. 1999: The stability of the thermohaline circulation in global warming experiments. *Journal of Climate* 12, 1117–33.

Senior, C.A. and Mitchell, J.F.B. 1993: Carbon-dioxide and climate – the impact of parameterisation. *Journal of Climate* 6, 393–418.

Spencer R.W. and Braswell, W.D. 1997: How dry is the tropical free troposphere?

Implications for global warming theory. *Bulletin of the American Meteorological Society* 78, 1097–106.

Trenberth, K.E. (ed.) 1992: *Climate system modelling*. Cambridge: Cambridge University Press.

Wilby, R.L. and Wigley, T.M.L. 1997: Downscaling general circulation model output: a review of methods and limitations. *Progress in Physical Geography* 21, 530–48.

|7|

Indirect impacts and options for change

Introduction

So far the direct climatic effects of global warming have predominantly been discussed. The only indirect effect covered was sea level change. The results predicted by the climate models are used in a variety of ways to try and assess the level of impact on human society and on the biosphere. Chapter 4 has already discussed how past climate changes has affected humans but how susceptible is our present society to climate change? Has our technology made us impervious to change or are we more sensitive now to the stresses and strains that may be caused by global warming?

The scientific assessment of climate change, working group 1, of the intergovernmental panel on climate change, is just one of several working groups addressing different aspects of climate change. Working groups 2 and 3 look at socio-economic aspects of global warming. Working group 2 concentrates on the vulnerability of human and natural systems to climate change and the potential to adapt (Watson *et al.*, 1996). Working group 3 considers reduction of greenhouse gas emissions and mitigation strategies. The following chapter is based on that report (Bruce *et al.*, 1996). Socio-economic assessment of the impact of global warming relies on the results gleaned from climate model studies. It looks at the effects of climate change and attempts to place a damage estimate on them. The latter has an important influence on the decision-making process with regard to responding to global warming. This book is primarily concerned with physical science, so most of the discussion will relate to the use of the science to predict the wider effects rather than delving into economic decision making. None the less the science is intimately entwined with the economics. This chapter will describe some of the ethical issues and introduce some terms used in the wider environmental debate.

Socio-economic effects of global warming

Pearce *et al.* (1996) provide an overview of the various indirect impacts of global warming. These are then subdivided into whether or not they have a market impact. Most of the studies presented in Pearce *et al.* are based on the results from equilibrium studies from AGCMs using instantaneous doubling of carbon dioxide. Therefore most of the socio-economic impact assessment is based on the 1990 IPCC which does not contain the transient model experiments or the effect of aerosols. Further uncertainties arise because different impact studies have used different GCMs. Some researchers have averaged their results to a standard 2.5 °C warming while others have not. All this in itself results in considerable uncertainty in the damage estimates (Pearce *et al.*, 1996). The following discussion concentrates on three high profile areas of indirect impacts of global warming: agriculture, health and ecosystems. The first two illustrate how climatic change will affect some of our basic needs and requirements. The other indicates how the natural world will be affected. They all show how, when we come to consider these sorts of impacts, we are swiftly confronted with ethical dilemmas. Even if the scientific predictions were certain, the way forward is far from clear.

Agriculture

Agriculture will be affected in two ways: first, by the carbon dioxide fertilization effect and second, the climate changes which will also affect growth and yield. Many of the changes concerning agriculture are related to temperature and precipitation. Not all of these will result in damage and some may aid crop growth. Decreases in crop yields will occur due to heat stress and decreased soil moisture, as well as related problems of soil erosion and drought. The pests and diseases that affect crops are also influenced by climate and this in turn will affect agriculture. Weeds can have a significant impact on crop yields and may benefit from the carbon dioxide fertilization effect. Many of the worse weeds are C4 species and will not benefit as much as the crops. There are cases where C3 weeds are associated with C4 crops (such as maize and millet) which may be detrimental to the crops (Rosenzweig and Hillel, 1998). There are benefits, however. Carbon dioxide fertilization and increased water use efficiency may well aid crop growth although this is a contentious issue. To fully exploit the carbon dioxide fertilization effect may require the use of additional nitrogen-based fertilizers which are not readily available in the developing world (Rosenzweig and Hillel, 1998).

On a global scale the predictions are for crop yields to fall and prices to rise with a consequent loss of human life. Such a crude picture, however, fails to portray the complexity of the likely crop changes and economic impact. There will be winners and losers and the losers in production may not necessarily be the losers economically. Studies show that producers tend to gain and consumers lose out due to increased prices (Rosenzweig and Hillel, 1998). The European Union may well experience increased agricultural yields due to the increases in temperature and precipitation with most of that gain occurring in the north (CRU/ERL, 1992). In contrast the United States may experience decreased yields but being a net exporter of grain US farmers would gain. This is because the decrease in yields would be more than offset by the increase in price (Kane *et al.*, 1992). In England and Wales it has been suggested that global warming will not affect the overall area of crop production but will significantly affect the pattern of production. For instance, by 2060 agriculture in the North Midlands and Fens was predicted to become uneconomic without global warming but with rising temperatures these would become important areas for cereal production (Hossell *et al.*, 1996). From a review of national and regional studies Rosenzweig and Hillel (1998) consider that the effect of global warming on agriculture in the tropical regions is likely to be greater than in the temperate regions. This, however, has as much to do with economic and social conditions as it has with global warming. Poorer nations dominate the tropical regions and their precarious infrastructure, and the political and economic tensions mean they are less able to cope with changing climate. Agriculture has the potential to adapt to any climate change by using different crops, genetically engineered crops, or changing farming practices. It is unlikely, however, that such changes in themselves could negate the effects of global warming on agriculture (Pearce *et al.*, 1996). There are also the indirect effects of global warming to consider. An increase in sea level may well result in a salinization of ground water which will have effects on agriculture and the water supply. Similarly salinization may result from over-irrigation.

Health

Changes in human health will again result in winners and losers and will be due to a variety of direct and indirect effects. Mortality may well rise in the more developed nations as death from coronary heart disease and stroke increases at the extreme temperature ranges. Increases in mortality in Greater London have been noted during spells of hot weather (Haines, 1991). There appears to be a U-shaped relationship between temperature and mortality with more deaths occurring at extreme temperatures. In Europe more deaths occur during extreme cold weather than in hot weather (Kovats *et al.*, 1999). A rise in temperature is likely to lead to a decrease in the number of people killed by cold weather. While this may benefit countries which have high rates of mortality in cold weather, such as Britain, there will be an increase in heat-induced deaths. Pearce *et al.* (1996) conclude that this increase is likely to more than offset any decrease. However, Martens (1998) concludes that changes in the thermal regime may lead to a decrease in mortality.

Severe air pollution is linked to longer, warmer summers which are likely to increase with global warming. Air pollution affects those suffering from respiratory tract problems such as asthma, and severe attacks can lead to death. The developed nations may see an increase in communicable diseases. Some diseases such as malaria, plague and typhus, which at present are associated with the tropics, occurred in more temperate zones in the past. This is because the insects which transmit the disease are sensitive to temperature. Climatic change could increase both the spread and range of diseases that require an animal or plant to carry them (vector-borne infectious diseases). Therefore mosquito-borne diseases in Australia and America could well increase. Even western Europe may not be immune. Increasing global travel means that insects such as malaria-carrying mosquitoes can also travel. In hot weather they can survive in the more temperature regions. During the summer of 1994 six people in the Paris airport region were affected by malaria from infected mosquitoes that had travelled on planes (Guillet *et al.*, 1998).

As well as the direct effect of temperature changes, the indirect effects of global warming will also affect the health of the world. The ten countries most vulnerable to a rise in sea level include Bangladesh, Egypt, Pakistan, Indonesia and Thailand. All these countries have large and relatively poor populations; increased flooding will cause death directly and indirectly through disease (Haines, 1991). It is also likely that more impoverished nations and individuals will suffer the consequences of these changes disproportionately. Poorer sectors of the population lack the necessary natural, technical and social resources to tackle these climatic-induced health changes (Pearce *et al.*, 1996). While Europe may be immune from some of the worst effects of flooding there are still potential health dangers. Flooding increases the risk of communicable diseases, wherever it occurs and there are other hazards from sewage and toxic chemicals that may be released. Flooding has also been linked to increased incidence of depression in Europe (Kovats *et al.*, 1999). In contrast, drought may lead to more death through starvation. Global warming may lead to a shortage of clean water supplies in the developing world. It may also result in an increase in the number of waterborne diseases (Martens, 1998). Another less obvious impact of global warming on health is an increased incidence of food poisoning in countries with a poor food safety record. This includes Britain, where a good relationship has been shown between the outbreak of food poisoning and the temperature of the month previous to the incident (Bentham and Langford, 1995).

Working group 3 of the IPCC 1995 caught the media attention by trying to place a value on the loss of life due to this increased mortality (Pearce *et al.*, 1996). The working group were widely reported as attempting to assess the value of a human life. They were accused of saying that a life in the developing world was worth less than one in the developed world (Pearce, 1995). Some economists have put a value on a 'northern life' as ten times that of a 'southern life' (Fankhauser, 1995; Jäger and O'Riordan, 1996). The working group itself made the following comments: 'Attributing a monetary value to a "statistical life" is controversial and raises a number of difficult and theoretical issues. It is important to understand that what is valued is a change in the risk of death, not human life itself.' It goes on to say that 'The reality is that safety is not "beyond price"' (Pearce *et al.*, 1996). By this it means that we all make decisions which involve a trade-off between safety and other socio-economic considerations. Perhaps the easiest example of this is in buying a car. A car built like a tank would be extremely safe, but it would also be extremely expensive to buy and who wants to turn up at the shops in a tank? This raises a moral dilemma. Should we base our decisions on political and ethical considerations valuing all life equally or should we consider how much people are willing to spend to decrease the risk of death? David Pearce, lead author of this chapter of the IPCC report, was quoted as arguing that 'This is a matter of scientific correctness versus political correctness' (Pearce, 1995). The idea of a statistical life illustrates a fundamental split between theoretical perspectives. There are those that believe science is neutral, offering value-free advice. Others see science as part of wider social processes and conclude that science cannot be impartial. Jäger and O'Riordan (1996) use the notion of a statistical life to question the ability of science to be neutral, particularly once that science has political implications. They argue that there can be no neutral analysis of the costs and benefits of climate change. Furthermore, they propose that climate change science must be analysed in the light of the political process and wider policy issues, such as the **north–south** divide, population growth, human rights and poverty. This will be more fully explored in the next chapter.

Ecosystems

The studies regarding the effects of climate change indicate that there is increased risk of extinction and therefore a loss in biodiversity. These extinctions will result from changes in habitat, predator/prey relationships and physiological factors. It is believed that global warming will result in a poleward migration of forests. For example, Figure 7.1 shows what might be expected from an increase of 5 °C in mean annual temperature and a 10 per cent increase in annual rainfall over Europe. At present much of Europe is covered by deciduous woodland with the Mediterranean countries having temperate evergreen forests and woody plants. Figure 7.1 shows the major vegetation zones shifting northwards by about 1100km (Warrick *et al.*, 1990). It is difficult to predict what sort of shift may occur with global warming. Many species studies are based on the 'climate envelope approach'. The present-day species distribution is related to the climate of the region. As the climate changes so the species moves with spatial changes in climate. However, such studies do not reflect the interactions between species, which may be upset by global warming and do not fully take into account population dynamics (Davis *et al.*, 1998). Therefore it is perhaps unwise to think that whole ecosystem shifts will occur. Ecosystems do not migrate as a whole.

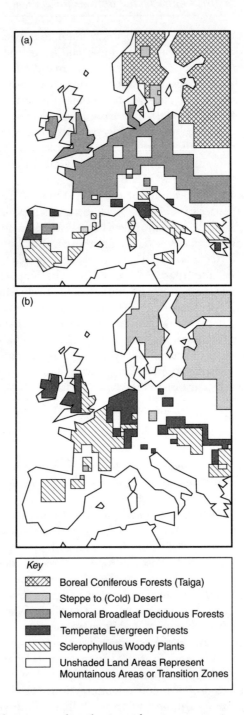

Figure 7.1 (a) The current distribution of major vegetation zones (b)Changes in the major vegetation zones caused by an increase of 5 °C in mean annual temperature and a 10 per cent increase in annual rainfall over Europe
Source: Warrick *et al.* (1990)

There are many conflicting factors, species type, spread of disease, soil moisture, acidity, etc. The other assumption commonly made in such studies is that the ecosystems are in equilibrium with the climate. Global warming will occur so quickly that changes in the ecosystem may lag significantly behind (several hundred years behind) the climate changes. Another factor is human management; ecosystems which move northward may find the land is not available because it has been given over to urban or agricultural development (Warrick *et al.*, 1990).

With some notable exceptions, climatic change does not necessarily lead to widespread species loss. However, when we start to consider conservation, in the socio-economic dimension, we find ethical dilemmas. Economists have to place a monetary value on something that does not have a value in traditional economics. This is to enable economists to perform a **cost benefit analysis** (CBA). CBA is a well-established economic tool for comparing the total costs of implementing a project to the benefits of that project. It differs from a financial appraisal in that it attempts to consider a project for society as a whole (Bateman, 1995; Griffin, 1998). As Griffin explains, the objective of CBA is to consider 'all humanly felt welfare changes'. This can lead to some environmental impacts being ignored. Also, it fails to encompass the concerns of those who feel that the advancement of human welfare is not the only goal to be achieved (Griffin, 1998). The problem for environmental economists is how to place a value where none has previously existed. Values can be placed on plants in terms of their current use in medicines, or stopping soil erosion (use value), or their potential future use (option value), or for just knowing they exist (existence value). The latter leads to estimates of how much the public would be **willing to pay** (WTP) to preserve endangered animals, species or ecosystems. This has resulted in a WTP value of $50 per person per year being put on the preservation of humpback whales (Pearce *et al.*, 1996). The idea of placing a value on an animal, or landscape, is not without its critics who see it as a step backwards, rather than forwards, in the protection of the environment (Bateman, 1995).

Response options

Increasing greenhouse gas emissions are a result of our modern industrial society. That industrial activity has also been a source of wealth to those nations that have successfully industrialized. The links between greenhouse gas emissions and economic activity are therefore very strong. Given this, how should society respond to increased greenhouse gas emissions and the potential threat of global warming? We could decide to do nothing at all about global warming, it is a perfectly valid response, particularly so if we believe the sceptics' view of global warming. To make vast changes to industry and agriculture would be pointless if the climate changes were negligible or did not occur at all. If we accept the global warming hypothesis, however, and reject the sceptics' view then we are still left with a decision about what should we do. In this section the general response options are considered rather than looking at individual policies. Responses to global warming are categorized in two ways. The first, called mitigation, is to pursue strategies which seek to lessen the greenhouse effect by reducing the amount of greenhouse gases in the atmosphere. The second is called adaptation which means seeking to reduce the negative impacts of global warming while capitalizing on the positive aspects (Riebsame *et al.*, 1995). However, within these two options it has to be recognized that changes in politics and economics occur on far shorter time-scales than global warming, which will take 50 to 100 years. This makes it very difficult to assess the success of the various response options as it will depend on the political and economic will to make them work (Jepma *et al.*, 1996).

Mitigation options

A lot of research has focused on the problem of mitigation. Again, two broad types emerge: those that deal with carbon dioxide reduction and those that investigate reducing the other greenhouse gases. Carbon dioxide emissions are principally related to either energy production or consumption (Jepma *et al.*, 1996). Therefore options concentrate on reducing the anthropogenic source of emissions or enhancing some of the sinks of carbon dioxide. Table 7.1 provides a summary of these options. The aim of both energy conservation and energy efficiency is to reduce the overall consumption of energy without changes to the method of production of energy. There is a subtle difference between conservation and efficiency. Conservation means turning off energy-consuming products when you don't need them; for instance, turning the television off completely rather than leaving it on standby. Efficiency means using products that require less energy to run. Examples of this are low energy light bulbs. Replacing standard bulbs with these will result in an energy saving and greater efficiency. Transport is another area where increased energy efficiency could have an impact. This is particularly important as the carbon dioxide emissions from transport are growing. In the EU it

Table 7.1 Measures for mitigating carbon dioxide.

Source reduction options	Sink enhancement options
Energy conservation and efficiency improvement	Capture and disposal of carbon carbon dioxide
Fossil fuel switching	Enhancing forest sink
Renewable energy	
Nuclear energy	

Source: After Jempa *et al.* (1996)

accounts for 26 per cent of carbon dioxide emissions (Smith and Owens, 1997). Hughes (1993) believes that the efficiency of the private car could be increased by up to 50 per cent if light weight cars were introduced. Energy efficiency is claimed to offer substantial reductions in carbon dioxide emissions. It is unclear, however, whether improvements in energy efficiency actually result in a reduction in overall energy consumption. This is particularly so at the household level where lowered energy bills may just result in the saving being spent on another energy-consuming device (Pearce, 1998).

The remainder of the options means turning to alternative ways of energy production. Fossil fuel switching requires using fossil fuels which have lower carbon dioxide emissions associated with them. For instance, coal contributes 44 per cent of all carbon dioxide emissions from fossil fuel combustion. Coal is the most inefficient fossil fuel in terms of the amount of heat released per unit emission of carbon dioxide emission. In contrast, natural gas is one of the most efficient heat sources with regards to carbon dioxide emissions (Grübler and Nakicenovic, 1994). Switching to carbon dioxide-free ways of producing electricity might seem an obvious way to reduce emissions. In 1971 an editorial in the journal *Nature* assumed that by the end of the twentieth century fossil fuels would have mostly been replaced by nuclear power. It failed to foresee the public mistrust of nuclear power. The accidents at Windscale, Three Mile Island and Chernobyl have further underlined the problems of nuclear power. This together with the economic constraints of decommissioning old nuclear power stations make this solution to global warming seem unlikely. Renewable energy sources also seem unlikely to be able to fill the gap (Collier and Löfstedt, 1997). Another solution is to capture and dispose of carbon dioxide from the energy production process before it is released. However, it is unclear how the captured carbon should be dealt with (Jepma *et al.*, 1996). The growing emission of carbon dioxide from transport could also be reduced by use of lower emission fuels.

Enhancing forest sinks have also received a great deal of attention. This requires new forest areas as existing mature forests are in equilibrium with the atmosphere. A problem with such sink enhancement programmes is that it is difficult to measure the amount of carbon sequestered. Claims have been made that the entire yearly carbon dioxide emissions of the USA are soaked up by the trees in North America (Fan *et al.*, 1998). Even if such a large sink could be substantiated, it would not last, particularly if the trees were harvested (Walker, 1998). Tropical forests are thought to sequester far more carbon that temperate forests. Therefore reforestation plans in tropical countries such as Malaysia and Brazil have been planned. There are worries, however, that such plantations are being used by the north's industry to avoid carbon dioxide emission reductions. Furthermore it has been calculated that it would require a tropical forest the size of India to soak up the required amount of carbon dioxide.

Adaptation options

Even if carbon dioxide emissions were held to present-day levels, because they are above pre-industrial levels the Earth is still committed to a proportion of global warming. Adaptation options work hand in hand with mitigation options. The less carbon dioxide put into the atmosphere, the smaller the consequent climate change and adaptation needed. Adaptation options try to achieve two goals. The first is to minimize the damage caused by global warming. The second is to enable societies and ecosystems to cope better with those facets of climate change that are inevitable (Jepma *et al.*, 1996). Perhaps, because human beings already adapt to climatic extremes, adaptation measures have received less attention than other options. In the face of global warming, however, current adaptation measures may not be enough and global warming is likely to increase the cost of these existing adaptations. There are three essential questions regarding adaptation. Which of the climate changes should we adapt to, how do we adapt to them and when should we adapt to them? Sea level rise provides an excellent example of the three types of adaptation measures commonly proffered: protection, retreat and accommodation. Sea level rise is difficult to estimate at a local level not just because of the uncertainties associated with climate change but also due to factors not associated with climate such as subsidence. Somerset in England provides two contrasting examples of how to manage sea level rise. At Porlock the decision was made to have a managed retreat in the face of sea level rise. Former farmland has been allowed to flood and form marshes. At Minehead, protection has occurred and over £12 million have been spent on increasing coastal defences. Sea level rise associated with global warming may mean that retreat will require more than just a loss in agriculture and forestry but also that people will have to migrate. Rahman and Huq (1998) estimate that over 94 million people who reside in the river delta regions of China, Egypt and Nigeria may be at risk from sea level rise.

Accommodation means that, like protection, more will have to be spent altering existing structures to a higher sea level but it will also involve a change of attitudes. More thought may have to be given to where buildings are constructed, avoiding the sea shore and known flood plains. Appropriate land use planning might ease the growing emissions from transport. New self-contained autonomous settlements with ease of access to public transport could reduce the need for private cars (Hughes, 1993; Smith and Owens, 1997). There would need to be an integrated transport policy. The experience of the Netherlands, where land use planning is most advanced, suggests that it will merely hold back the tide rather than reverse it. Unless fundamental assumptions about the link between economic growth and transport growth are challenged, nothing will change (Smith and Owens, 1997). Thus there is a wide range of areas which may require adaptation. For some areas of the globe such as the Pacific atolls, even all these measures may not be enough, because the predicted sea level increases will engulf these island nations, covering up to 80 per cent of their land surface (Howes *et al.*, 1997). There is also the problem of when to adapt. Adaptation can be proactive, that is, in advance of any climate change, or reactive, that is, after the event. Rosenzweig and Hillel (1998) consider that the recent down-grading of global warming estimates in the 1995 IPCC report (Houghton *et al.*, 1996) has led agricultural planners to adopt a '**wait-and-see**' approach. This may not be appropriate since waiting to adapt may ultimately result in higher costs.

Limiting the damage

Pearce *et al.* (1996) discuss the two main approaches that are used to analyse the decisions made in response to global warming and therefore to cost the damage. These are the cost-benefit framework and the sustainability approach. In the cost benefit analysis damage and adaptation costs are compared to the costs of mitigation. Using traditional CBA in this way leads to some interesting ethical issues. Apart from whether willingness to pay (WTP) in itself is right or not it can also lead to a distinction between rich and poor. High income groups, such as the developed nations, will have far more disposable income to save a whale than the low income developing nations. Therefore the choices of the rich will outweigh the poor. The extreme example of this is the 'statistical life' mentioned earlier. How should this issue be addressed? Should the developed world dictate the important issues or should the different income groups be weighted accordingly? Another problem is how to deal with future generations and their wishes. What will our children want and will it be the same as our desires? After all, this is to save the world, not for us, but for our children and grandchildren. CBA needs to take account of passing time. The past is seen as unchangeable so any costs spent then are seen as 'sunk' and not affecting the future. To costs and benefits in the future a '**discount rate**' is applied to ascertain their present day value. The idea of a discount rate comes from the premise that a penny saved today could be worth two tomorrow. In business this can be seen in terms of investment in equipment or training leading to greater productivity and profits. For greenhouse control it may mean that a few thousand dollars spent today will avoid millions of dollars of damage in the future (Arrow *et al.*, 1996). It all depends on the discount rate used. If the rate is set too high, then too little will be spent on avoiding climate change. If the rate is too low, then money spent on climate change is wasted and could have been spent on other needs. A garden left unattended for five years will be overgrown with weeds and the time and expense needed to weed the garden will be large. If instead a little bit of time and money is spent every year weeding the garden, then in five years' time there will still be weeds but the problem will not be large. The question is, how much do we invest every year tending our garden?

Discount rates are highly problematical requiring both a time horizon and a rate to be set. The longer the time horizon the more difficult it becomes to assess the benefits and costs. This is particularly so if the costs and benefits accrue over different time periods (Swaney, 1997). Munasinghe *et al.* (1996) emphasize the importance of discount rates saying that if the correct discount rates are not used inefficient policy will result. Arrow *et al.* (1996) describe the two main opposing methodologies for estimating discount rates: the prescriptive and descriptive approaches. The prescriptive approach takes an ethical viewpoint; the only way to transfer any compensation to future generations is through instigating climate change policies. This approach leads to lower discount rates and tends to favour more spending on climate change mitigation. In contrast the descriptive approach suggests maximizing consumption such that future generations receive the maximum economic benefit leaving them to decide how to spend it. The descriptive approach offers higher discount rates and therefore suggests less spending on mitigating climate change. The cost-benefit framework need not be done exclusively in monetary terms. Other objectives can be used although, as yet, no clear set of objectives have been agreed upon (Munasinghe *et al.*, 1996). This is called a multicriteria analysis (MCA). At its heart there still lies CBA, but it attempts to quantify other objectives that should be 'traded off' against the economic efficiency. Thus policy-makers can look to maximize a number of objectives.

The sustainability viewpoint of how society should address the global environmental challenge, of which global warming is the ultimate example, has become widely espoused of late. It is perceived that society will transform itself (development) because it has to due to the environmental pressures upon it (sustainability). There is no satisfactory definition of **sustainable development**. This is probably because of its rather amorphous nature. According to Blowers (1997) it can be a scientific principle, a political goal, a social practice and a moral guide all at the same time. Goldemberg *et al.* (1996) offer the widely quoted Bruntland Commission definition: 'sustainable development is development that meets the needs of the present without compromising the ability of future generations to meet their own needs' (World Commission on Environment and Development, 1987). It differs most significantly from the CBA viewpoint in that future generations are most important and harm to them should be avoided. Benefits and costs cannot be traded off (Pearce *et al.*, 1996). Even sustainability, however, is open to a variety of economic interpretations. Some approaches, termed weak sustainability, argue that the depletion of natural resources is all right so long as future generations can achieve the same standard of living as the present generation. Others argue for strong sustainability, that there can be no substitutes for some natural resources such as the atmosphere or climate (Goldemberg *et al.*, 1996). Supporters of the sustainability approach argue that there is enough evidence to suggest that damage will be done to future generations by increasing greenhouse gases. So how do we avoid or minimize that harm? Do we change our ways despite the costs or do we do as much as we can without 'unacceptable cost' to ourselves? Sustainability gives rise to another popular term called the '**precautionary principle**'. This means that someone seeking to introduce a new product or process should be able to prove that it will not harm the environment or at least make some sort of financial provision to put things right if it does (O'Riordan, 1995). The precautionary principle is open to a wide variety of interpretations depending on the context in which the term sustainabilty is being used (O'Riordan and Cameron, 1994). Sustainability often has as an objective that the global annual average surface temperature should not increase by a certain amount, whatever the cost (Pearce *et al.*, 1996). The problem for developing countries is the cost maybe lowered economic growth (Goldemberg *et al.*, 1996).

Equity

Rowbotham (1996) considers that the United Nations Framework Convention on Climate Change (UNFCCC) recognized seven basic principles:

1. Greenhouse gas emissions need to be reduced.
2. Developed countries need to take the lead in reduction.
3. The worries of the developing nations are legitimate.
4. Some countries are particularly vulnerable to the impacts of global warming.
5. The economies of some countries are heavily reliant on fossil fuel production.
6. That measures to address global warming must be cost–effective.
7. That the precautionary principle must be applied.

Turning these principles into legislation which nations can implement to control greenhouse gas emissions is a very difficult task. Banuri *et al.* (1996) identify five ways in which countries differ in their ability to respond to climate change. These are wealth (and consumption), emissions, impacts, national characteristics and natural resources. On studying these factors it becomes obvious that most are divided by wealth. This means there is a division between the developed and developing world (north and south). Over half the world's population live in the 42 poorest countries; most of their greenhouse emissions are due to subsistence activity such as cooking and keeping warm. It has been calculated that the developing world has accounted for only 16.2 per cent of past carbon dioxide emissions from industrial sources and just under a third of all carbon dioxide emissions (*see* Fig. 7.2). Their current emissions remain small (*see* Fig. 7.3). There is a factor of eight in the difference in carbon dioxide per capita emissions between the developed and developing world. North America, which has the highest per capita emissions, has 25 times the emissions of India and Asia (excluding China) which have the lowest per capita emissions (Grübler and Nakicenovic, 1994). The developing nations are likely to be the hardest hit by the impacts yet have the least capacity to cope. In contrast, in the developed world most greenhouse emissions come from the use of energy-guzzling appliances such as cars. The developed nations dominate past and present emissions of carbon dioxide (Figures 7.2 and 7.3). The developed world is likely to be the least affected and yet the best able to cope with climate change.

These emissions figures are open to dispute and many are subject to considerable uncertainty (Subak, 1996). This has led to a north–south divide when it comes to promoting priorities. The north tend to concentrate on environmental protection and cost effectiveness. The south concentrate on developmental priorities and historical responsibilities (Rowbotham, 1996). Overcoming these inequities and allocating future emissions and costs pose the greatest challenge to tackling global warming. Grübler and Nakicenovic (1994) provide a comprehensive analysis of past and present emissions, together with a consideration of how allocating future reductions might occur. They consider four ways to reduce greenhouse gas emissions to 4Gt by 2050. The first was to consider equal per capita emissions. This was the only scenario which allowed developing countries the chance to increase emissions. The developed world, however, faced substantial cuts with North America required to reduce emissions to 7 per cent of current levels by 2050. This level would certainly not be accepted. It has also been argued that such an approach rewards population growth (Banuri et al., 1996). The second method considered was equal percentage cuts from current emissions to desired target levels. The results were very sensitive to the reference year chosen and it all depended on which emissions were included. For example emissions from the biosphere, particularly those related to deforestation, are very uncertain but are of great importance to the developing world. In the case of Brazil the difference between taking a high estimate or a low estimate for biosphere emissions could mean changing its emissions by a factor of 10 (Grübler and Nakicenovic, 1994). Emissions of methane are also largely from the south due to the greater proportion of livestock, animal waste and paddy fields. Therefore it would be in the interests of the developing world to promote scenarios which include only industrial carbon in the emission assessment. The third method discussed cutbacks proportional to past regional emissions. Although all regions would be affected, significant cuts were again required by the north. Finally, equal per capita emissions equivalent to natural greenhouse gas sinks were considered. This approach produced impossibly low greenhouse gas allocations (Grübler and Nakicenovic, 1994).

As Banuri et al. (1996) point out, there are a bewildering array of emissions allocations proposals. All the proposals try to be 'fair' but there are difficult decisions to be made and major hurdles to overcome. All the allocations will require someone to pay for them and most likely this means a transfer of funds from the developed to the developing world. This will mean addressing questions not just about historical responsibility for greenhouse gas emissions and ability to pay, but questions about global poverty, trade and debt responsibilities. Jäger and O'Riordan (1996) point out that many Third

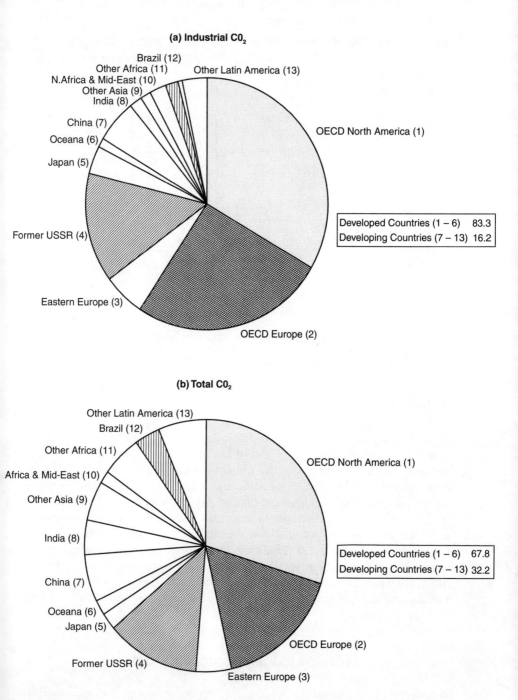

(a) Industrial CO₂

Brazil (12)
Other Africa (11)
N.Africa & Mid-East (10)
Other Asia (9)
India (8)
China (7)
Oceana (6)
Japan (5)
Former USSR (4)
Eastern Europe (3)
Other Latin America (13)
OECD North America (1)
OECD Europe (2)

Developed Countries (1 – 6)   83.3
Developing Countries (7 – 13) 16.2

(b) Total CO₂

Other Latin America (13)
Brazil (12)
Other Africa (11)
Africa & Mid-East (10)
Other Asia (9)
India (8)
China (7)
Oceana (6)
Japan (5)
Former USSR (4)
Eastern Europe (3)
OECD North America (1)
OECD Europe (2)

Developed Countries (1 – 6)   67.8
Developing Countries (7 – 13) 32.2

Figure 7.2 Greenhouse gas emissions broken down by region for (a) industrial carbon dioxide emissions and (b) total greenhouse gas emissions
Source: Grübler and Nakicenovic (1994)

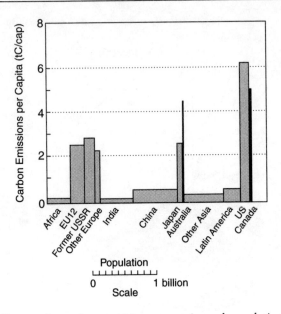

Figure 7.3 Present-day carbon emissions per capita and population in 1993
Note: The height of each block indicates the per capita emissions of carbon dioxide from all sources of fossil fuel consumption. The width of each block indicates the population and the area of the block is proportional to total fossil carbon dioxide emissions
Source: Bruce *et al.* (1996)

World countries would argue that deforestation, and the consequent greenhouse gas emissions, are not of choice but of necessity. They argue that the necessity is due to the north's imposition of unfair trade agreements and large debt repayments. The developed world has industrialized at an environmental cost, increased greenhouse gases is just one of them. Industrialization has long been perceived as the route to wealth and power. It is quite understandable that much of the world still wishes to develop into full industrialized states. However, as Clayton (1995) says 'A molecule of greenhouse gas emitted anywhere becomes everyone's business'. Many of the arguments at Kyoto in December 1997, which set greenhouse emission targets, and the succeeding meetings to sort out how to reach those targets, centre on how to keep all the nations committed.

Integrated assessment models

Whether a CBA or sustainability approach is adopted we need to know which response options will work best and, what effects they will have on economic activity and in turn, what will then happen to emissions and the climate. **Integrated Assessment Models** (IAMs) are employed to research these important policy options and use the science of climate change directly. Integrated assessments are an attempt to link together the various disciplines that study global warming. IPCC working group 3 (Weyant *et al.*, 1996) use this as their definition of an IAM. They point out that while the levels of detail in each model vary, they are all interdisciplinary, combining knowledge from several research areas. The aim of these studies is to explore how both human society and the natural world may react to global warming, to consider the policy options available and to assess what research would help identify appropriate policy responses. However, IAMs can also provide an important contribution to the scientific debate. This research is done by linking together various computer-based models, which are mathematical representations, from the relevant disciplines. Weyant *et al.* (1996) note that this ensures consistency between the assumptions in the models but limits the type of information represented in the model.

'Full-scale' IAMs attempt to consider the full range of issues presented by global warming. A full-scale IAM will not only include all the principal concerns but also the linkages between them. Figure 7.4 illustrates the highly dependent nature of the systems to be modelled and shows the many components that have to be modelled. Figure 7.4 shows that the components range from a climate model to models of human activities such as agriculture and health. As McGuffie and Henderson-Sellers (1997) note, some of the philosophies that underpin climate modelling have their parallels in IAMs. EBMs and GCMs both attempt to model the climate; the former from a set of simple general equations, the latter from a set of complex equations derived from first principles. IAMs also follow this outline with some models

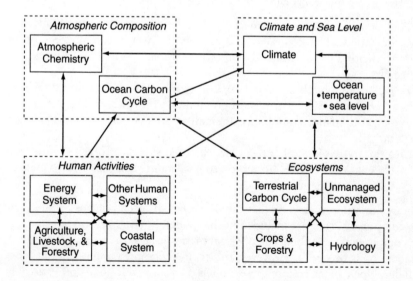

Figure 7.4 The main components needed for a full integrated assessment model
Source: Bruce *et al.* (1996)

taking a simple general approach and others attempting to model all the processes. Furthermore, as with climate models some IAMs focus on global predictions while others make small scale regional assessments. There are other 'characterizations' of IAMs that can be used such as how they treat uncertainty and decision-making and which impacts they attempt to assess. There are two major design considerations for IAMs. First, how to integrate the different models and second, how to deal with the large number of uncertainties that the modellers face (Weyant *et al.*, 1996). Integrating the different types of models has proved highly difficult. The uncertainties pervade practically every part of the model, from how much emissions are likely to grow, to the future of populations and economies. Many early impact studies dealt with this by taking a 'best estimate'. Now, however, uncertainty is increasingly incorporated into models. This allows both well-known risks, uncertainty and surprise events to be analysed (Dowlatabadi and Morgan, 1993; Rosenzweig and Hillel, 1998).

Weyant *et al.* (1996) consider two types of IAMs: policy optimization models and policy evaluation models. Policy optimization models seek to optimize variables which are subject to policy control, such as carbon taxes, given that there are certain policy goals, such as minimizing the cost to industry. These models concentrate on the socio-economic detail and provide a lot of information about the losses and gains associated with following different courses of action. Cost benefit analysis is at the heart of many of these models. To make a cost benefit analysis in policy optimization models the details of the physical processes are minimized, with economic losses often simply related to the average global surface temperature. Weyant *et al.* (1996) further sub-divide this model type into three categories.

1. Cost benefit models are driven by cost benefit analysis and economic optimization. The goal of these models is to ensure that the cost of implementing mitigation and adaptation strategies does not outweigh the benefits.
2. Target-based models are also based on cost benefit analysis but they have to achieve their targets given certain guiding principles. Thus their goal is to make the cost of compliance as small as possible while still achieving a set target.
3. Uncertainty-based models recognize that many of the processes within the cost benefit type models are subject to uncertainty and they seek to include that directly in the model.

Policy evaluation models predict the outcomes of given policies on the various components that are included in the model. In policy evaluation models the components describing the physical processes are complex and contain a great amount of detail. However, the socio-economic components are less well developed and therefore these models do not provide good cost estimates for the impacts of climate change. These types of model can be split into two sub-categories (Weyant *et al.*, 1996).

1. Deterministic projection models where each input and output take on a single value.
2. Stochastic projection models where some of the inputs and outputs are treated as being random.

Although IAMs include climate models and estimates of emissions these are not the same as discussed in the previous chapters. For instance, full-scale IAMs must consider the emissions of greenhouse-related gases from both human activities and natural ecosystems. Thus although there is a great deal of overlap, the gases included in IAMs are slightly different from those in climate GCMs (Weyant *et al.*, 1996). Whereas GCMs include water vapour, ozone, carbon dioxide, methane, nitrous oxide, sulphur aerosols and CFCs, IAMs ignore water vapour and ozone because they are not emitted in quantities large enough to be important. They include, however, important precursors to ozone production such as carbon monoxide, nitrogen oxides and non-methane hydrocarbons. Similarly, it is the emission of sulphur oxides, which determine the amount of sulphate aerosols rather than the aerosols themselves, that are considered in IAMs. From the emissions, the climatic components such as temperature rise can be calculated. For example, in PAGE95, a policy evaluation model, a relatively simple set of equations are used to calculate the global and regional temperature rise from emissions (Plambeck *et al.*, 1997). The level of climate sensitivity is set using GCM results and this is the sort of parameter that is highly uncertain. The PAGE95 model incorporates this uncertainty by running the model many times using a random sampling technique for the value of the sensitivity and other uncertain parameters. The results are then a probability function for each model output.

Until 1992 there had only been two IAMs, now there are 20 and the total looks set to increase (Weyant *et al.*, 1996). IAMs are in a process of continual development and there is no single methodology. Many of the results in the 1995 IPCC Working Group 3 report (Bruce *et al.*, 1996) are based on equilibrium model results. The inclusion of sulphate aerosol cooling and changes in the estimation of global warming potentials of fluorocarbons all require IAMs to be updated. Plambeck *et al.* (1997) include such recent advances in the understanding of climate change. Of particular note is that PAGE95 predicts that the level of damages in the developing world are greatest because of their dependence on economic activities which are most affected by climate change coupled with the lack of resources to adapt (Plambeck *et al.*, 1997). It would be wrong, however, to think that these models only contribute to the social science debate. Policy evaluation models have also illuminated the scientific debate. They have been used to investigate the carbon cycle, global warming potentials, sulphate aerosols and the effect of delaying any response to global warming (Weyant *et al.*, 1996).

It is in this later category that some of the most hotly debated results have arisen and have provided another source of uncertainty. Studies with a policy evaluation model called IMAGE 1.0 (Integrated Model to Assess the Greenhouse Effect) assessed the impact of delaying response to global warming. In 1990 it was found that if greenhouse gases remained uncontrolled for a further 10 years, until 2000, then it would result in only a small increase in global average surface temperature. However, it would then require a 20 per cent reduction in carbon dioxide emissions from the year 2000 levels compared to a 5 per cent reduction if control policies were implemented in 1990 (Rotmans, 1990). Thus the analysis showed that the timing of implementing control policies is important in affecting climate change. It also showed that there is limit to how long control policies can be delayed because there will come a point when the amount of reduction in emissions required will be too great to be implemented. This is echoed by the work of Hasselmann *et al.* (1997) with SIAM (Structural Integrated Assessment Model) they also concluded that although the long time horizons involved in global warming allow for flexibility, delaying mitigation only increases the cost.

Similar studies have also been completed with policy optimization models, which attempt to put a cost on changes. These studies have concluded that it would be cheaper to initially reduce greenhouse gas emissions by a small amount than to have substantial reductions later on. It would be more expensive to have large reductions in the short term (Kosobud *et al.*, 1994; Richels and Edmonds, 1995). It has been argued that the reason for this is that we have a budget, in this case carbon, that has to be invested over a set period of time. The question is how best to invest that budget. Economists would argue that it is best to spend (burn carbon) now rather than later (Wigley *et al.*, 1996). Weyant *et al.* (1996) have pointed out that such studies should not be seen to support a 'do nothing or wait and see approach' and that the studies assume a commitment to reduce greenhouse emissions both by using current technology and investment in research and development to produce technologies which reduce emissions. As might be expected, however, this is not how the studies have been treated. Accusations have been made that these are excuses for doing nothing and support the sceptic line on global warming (Pearce, 1996).

It should be remembered that just as GCM results reflect the type of parameterizations used so IAM results will reflect the choice of discount rates and the other choices made. IAM results cannot be assessed without considering the in-built theoretical approaches of the modellers. There are many subjective judgements that have to be made and although based on scientific rationale they cannot be empirically tested. This results in widely differing cost estimates as to the damage caused to economies by global warming (Howarth and Monahan, 1996). For example, Azar and Sterner (1996) consider the work of Nordhaus who has produced several IAMs most notably DICE and RICE which are essentially CBA policy optimization models (Nordhaus, 1993; Nordhaus and Yang, 1996). Studies with the earlier DICE model led to the conclusion that only modest reductions in greenhouse gas emissions were justified economically (Nordhaus 1993). Azar and Sterner (1996) considered three aspects of the model: the carbon cycle model, discount rate and the inequalities between north and south incomes. By varying these they achieved cost estimates 50 to 100 times higher than Nordhaus (1993) leading Azar and Sterner (1996) to conclude that large reductions in carbon dioxide levels were justified on purely economic grounds.

Summary

This chapter would seem to suggest that even our technologically advanced society is not immune to the changes that will be brought about by global warming. It is clear that a lot of research needs to be done, not only in improving our knowledge of how the climate might change but also, with regards to the cost and benefits to society of the various possible responses to global warming. Our advantage over our ancestors, who also had to deal with climate change, is that we can, to an extent, predict the changes and the consequences. For many the indirect consequences of global warming, such as sea level rise, would appear to be a greater threat than the direct consequence of a temperature change. Some nations are in danger of disappearing completely! This chapter has considered some of the options available to us to stop, or at least slow down, the effects of global warming. It has looked at the use of the science of global warming in a wider arena. While at times the controversy surrounding these options has been mentioned, the real difficulties of implementation have been left undiscussed. Some of these problems relate to how you introduce measures to encourage people to reduce their fossil fuel consumption. This requires detailed consideration of the relative merits of, say, legislative control versus voluntary agreements. This book will not cover these issues, because, as the next chapter will briefly discuss, individual nations are likely to take very different policy decisions. The next chapter will look at how the science of global warming became a political issue and the role scientists played in that politicization. This perhaps helps to explain why global warming remains such a contested issue despite the large degree of scientific consensus and the apparently severe consequences of inaction.

References

Arrow, K.J., Cline, W.R., Maler, K.-G., Munasighe, M., Squitieri, R. and Stiglitz, J.E. 1996: Intemporal equity, discounting and economic efficiency. In Bruce, J.P., Lee, H. and Haites, E.F. (eds), *Climate change 1995: economic and social dimensions of climate change.* Cambridge: Cambridge University Press.

Azar, C. and Sterner, T. 1996: Discounting and distributional considerations of global warming. *Ecological Economics* **19**, 169–84.

Banuri, T., Göran-Mäler, K., Grubb, M., Jacobson, H.K. and Yamin, F. 1996: Equity and social considerations. In Bruce, J.P., Lee, H. and Haites, E.F. (eds), *Climate change 1995: economic and social dimensions of climate change.* Cambridge: Cambridge University Press.

Bateman, I. 1995: Environmental and economic appraisal. In O'Riordan, T. (ed.), *Environmental science for environmental management.* Harlow: Longman Group.

Bentham, G. and Langford, I.H. 1995: Climate change and the incidence of food poisoning in England and Wales. *International Journal of Biometeorology* 39, 81–6.

Blowers, A. 1997: Environmental policy: ecological modernisation or the risk society? *Urban Studies* 34, 845–71.

Bruce, J.P., Lee, H. and Haites, E.F. (eds) 1996: *Climate change 1995: economic and social dimensions of climate change.* Cambridge: Cambridge University Press.

Clayton, K. 1995: The threat of global warming. In O'Riordan, T. (ed.), *Environmental science for environmental management.* Harlow: Longman Group.

Collier, U. and Löfstedt, R. E. 1997: Comparative analysis and conclusions. In Collier, U. and Löfstedt, R. E. (eds). *Cases in climate change policy: political reality in the European Union.* London: Earthscan.

CRU/ERL 1992: *Development of a framework for the evaluation of policy options to deal with the greenhouse effect: economic evaluation of impacts and adaptive measures in the European Communities.* Climate Research Unit (CRU). London: University of East Anglia and Environmental Resources Limited (ERL).

Davis, A.J., Jenkinson, L.S., Lawton, J.H., Shorrocks, B. and Wood, S. 1998: Making mistakes when predicting shifts in species range in response to global warming. *Nature* 391, 783–6.

Dowlatabadi, H. and Morgan, M.G. 1993: Integrated assessment of climate change. *Science* 259, 1813, 1932.

Fan, S., Gloor, M., Mahlonan, J., Pacala, S., Sarmiento, J., Takahasi, T. and Tans, P. 1998: A large terrestrial carbon sink in North America implied by atmospheric and oceanic carbon dioxide data and models. *Science* 282, 442–6.

Fankhauser, S. 1995: *Valuing climate change: the economics of the greenhouse effect.* London: Earthscan.

Goldemberg, J., Squitieri, R., Stiglitz, J., Amano, A., Shaoxiong, X. and Saha, R. 1996: Introduction: scope of the assessment. In Bruce, J.P., Lee, H. and Haites, E.F. (eds), *Climate change 1995: economic and social dimensions of climate change.* Cambridge: Cambridge University Press.

Griffin, R.C. 1998: The fundamental principles of cost-benefit analysis. *Water Resources Research* 34, 2063–71.

Grübler, A. and Nakicenovic, N. 1994: *International burden sharing in greenhouse gas reduction.* RR-94-9, June 1994. Laxenburg: International Institute for Applied Systems Analysis.

Guillet, P., Germain, M.C., Giacomini, T., Chandre, F., Akogbeto, M., Faye, O., Kone, A., Manga, L. and Mouchet, J. 1998: Origin and prevention of airport malaria in France. *Tropical Medicine and International Health* 3, 700–5.

Haines, A. 1991: Global warming and health. *British Medical Journal* 302, 669–70.

Hasselmann, K., Hasselmann, S., Giering, R., Ocana, V. and VonStorch, H. 1997: Sensitivity study of optimal CO_2 emission paths using a simplified structural integrated assessment model (SIAM). *Climatic Change* 37, 345–86.

Hossell, J.E., Jones, P.J., Marsh, J.S., Parry, M.L., Rehman, T. and Tranter, R.B. 1996: The likely effects of climate change on agricultural land use in England and Wales. *Geoforum* 27, 149–57.

Houghton, J.T., Meira Filho, L.G., Callander, B.A., Harris, N., Kattenberg, A. and Maskell, K. (eds) 1996: Climate change 1995: *The science of climate change*. Cambridge University Press, Cambridge.

Howarth R.B. and Monahan, P.A. 1996: Economics, ethics and climate policy: framing the debate. *Global and Planetary Change* 11, 187–99.

Howes, R., Skea, J. and Whelan, B. 1997: *Clean and competitive: motivating environmental performance in industry*. London: Earthscan.

Hughes, P. 1993: *Personal transport and the greenhouse effect*. London: Earthscan.

Jäger, J. and O'Riordan, T. 1996: The history of climate change science and politics. In O'Riordan, T. and Jäger, J. (eds), *Politics of climate change: a European perspective*. London: Routledge.

Jepma, C.J., Asaduzzaman, M., Mintzer, I., Maya, R.S. and Al-Moneef, M. 1996: A generic assessment of response options. In Bruce, J.P., Lee, H. and Haites, E.F. (eds), *Climate change 1995: economic and social dimensions of climate change*. Cambridge: Cambridge University Press.

Kane, S., Reilly, J. and Toeby, J. 1992: An empirical study of the economic effects of climate change on world agriculture. *Climatic Change* 21, 17–35.

Kosobud, R., Daly, T., South, D. and Quinn, K. 1994: Tradeable cumulative CO_2 permits and global warming. *The Energy Journal* 15, 213–32.

Kovats, R.S., Haines, A., Stanwell-Smith, R., Martens, P., Menne, B. and Bertollini, R. 1999: Climate change and human health in Europe. *British Medical Journal* 318, 1682–5.

Martens, P. 1998: *Health and climate change: modelling the impacts of global warming and ozone depletion*. London: Earthscan.

McGuffie, K. and Henderson-Sellers, A. 1997: A climate modelling primer. 2nd edition. Chichester: John Wiley and Sons.

Munasinghe, M., Meier, P., Hoel, M., Hong, S.W. and Aaheim, A. 1996: Applicability of cost-benefit analysis to climate change. In Bruce, J.P., Lee, H. and Haites, E.F. (eds) *Climate change 1995: economic and social dimensions of climate change*. Cambridge: Cambridge University Press.

Nature (editorial) 1971: The great greenhouse scare. *Nature* 229, 514.

Nordhaus, W.D. 1993: Rolling the DICE: an optimal transition path for controlling greenhouse gases. *Resource and Energy Economics* 15, 27–50.

Nordhaus, W.D. and Yang Z. 1996: A regional dynamic general equilibrium model of alternative climate-change strategies. *The American Economic Review* **86**, 741–65.

O'Riordan, T. 1995: Environmental science on the move. In O'Riordan, T. (ed.), *Environmental science for environmental management*. Harlow: Longman Group.

O'Riordan, T. and Cameron, J. 1994: The history and contemporary significance of the precautionary principle. In O'Riordan, T. and Cameron J. (eds), *Interpreting the precautionary principle*. London: Earthscan.

Pearce, D.W., Cline, W.R., Achanta, A.N., Fankhauser, S., Pachauri, R.K., Tol, R.S.J. and Vellinga, P. 1996: The social costs of climate change: Greenhouse damage and the benefits of control. In Bruce, J.P., Lee, H. and Haites, E.F. (eds), *Climate change 1995: economic and social dimensions of climate change*. Cambridge: Cambridge University Press.

Pearce, F. 1995: Global row over value of human life. *New Scientist* **147** (no. 1991), 7.

Pearce, F. 1996: Sit tight for 30 years, argues climate guru. *New Scientist* **149** (no. 2013), 7.

Pearce, F. 1998: Consuming myths. *New Scientist* **159** (no. 2150), 18–19.

Plambeck, E.L., Hope, C. and Anderson, J. 1997: The PAGE95 model: integrating the science and economics of global warming. *Energy Economics* **19**, 77–101.

Rahman, A. and Huq, S. 1998: Coastal zones and oceans. In Rayner, S. and Malone, E.L. *Human choices and climate change volume 2: resources and technology*. Columbus, OH: Batelle Press.

Richels, R. and Edmonds, J. 1995: The cost of stabilizing atmospheric CO_2 concentrations. *Energy Policy* **23**, 373–8.

Riebsame, W.E., Strzepek, K.M., Westcoat, Jr., J.L., Gaile, G.L., Jacobs, J., Leichenko, R., Magadza, C., Perritt, R., Phien, H., Urbiztondo, B.J., Restrepo, P., Rose, W.R., Saleh, M., Tucci, C., Li, L.H. and Yates, D. 1995: Complex river basins. In Strzepek, K.M. and Smith, J.B. (eds), *As climate changes: international impacts and implications*. Cambridge: Cambridge University Press.

Rosenzweig, C. and Hillel, D. 1998: *Climate change and the global harvest*. Oxford: Oxford University Press.

Rotmans, J. 1990: IMAGE: *An integrated model to assess the greenhouse effect*. Dordrecht: Kluwer.

Rowbotham, E.J. 1996: Legal obligations and uncertainties in the climate change convention. In O'Riordan, T. and Jäger, J. (eds), *Politics of climate change: a European perspective*. London: Routledge.

Smith, J. and Owens, S. 1997: Routes of reducing transport-related greenhouse gas emissions. In Collier, U. and Löfstedt, R. E. (eds), *Cases in climate change policy: political reality in the European Union*. London: Earthscan.

Subak, S. 1996: The science and politics of national greenhouse gas inventories. In O'Riordan, T. and Jäger, J. (eds), *Politics of climate change: a European perspective*. London: Routledge.

Swaney, J.A. 1997: The basic economics of risk analysis. In Molak, V. (ed.), *Fundamentals of risk analysis and risk management*. Boca Raton: Lewis Publishers.

Walker, G. 1998: A perfect excuse: if America's trees soak up its carbon, climate deals may be off. *New Scientist* **160** (no. 2157), 5.

Warrick, R.A., Barrow, E.M. and Wigley, T.M.L. 1990: *The greenhouse effect and its implications for the European Community*. Luxembourg: Commission of European Communities.

Watson, R.T., Zinyowera, M.C. and Moss, R.H. 1996: *Climate change 1995: impacts, adaptations and mitigation of climate change: scientific and technical analyses*. Cambridge: Cambridge University Press.

Weyant, J., Davidson, O., Dowlatabadi, H., Edmonds, J., Grubb, M., Pearson, E.A., Richels, R., Rotmans, J., Shukla, P.R., Tol, R.S.J., Cline, W. and Frankhauser, S. 1996: Integrated assessment of climate change: an overview and comparison of approaches and results. In Bruce, J.P., Lee, H. and Haites, E.F. (eds), *Climate change 1995: economic and social dimensions of climate change*. Cambridge: Cambridge University Press.

Wigley, T., Richels R. and Edmonds, J. 1996: Economic and environmental choices in the stabilization of atmospheric CO_2 concentrations. *Nature* **379**, 240–3.

World Commission on Environment and Development 1987: *Our common future*. London: Oxford University Press.

8

The need for consensus

Introduction

The previous chapter introduced some of the ways in which the physical sciences are used to inform policy decisions and this raised some interesting ethical issues. It would be impossible for this book to address all these issues and many of the them pertain to wider problems such as the relationship between the north and south; the developed and developing world. This book is looking at the science of climate change and this chapter is devoted to the way science has informed the wider issues that surround global warming. It will also consider how it has been received by those in the developed nations, arguably the nations with the most responsibility to respond.

One problem that haunts global warming is the need for scientific certainty. This chapter will consider this aspect as many still call for certainty as a prerequisite for action. As Howes *et al.* (1997) relate, the fossil fuel industry, the oil-producing states and an alliance of Japan, USA, Canada, Australia and New Zealand (JUSCANZ) used this uncertainty to delay action at the CoP-1 meeting in Berlin in 1995. This ended any hope the Association of Small Island States (AoSIS) had for a 20 per cent reduction in emissions by 2005. Therefore despite the widespread scientific view that global warming is a major problem, the views of the sceptics remain as an issue to be addressed. This is because the sceptics' view, that global warming at worse will be insignificant, support economic arguments for inaction. The scientific evidence has become a battleground, and although the sceptics may have lost the battle it is as yet unclear whether they will lose the war. While academic texts may exhort the need for change, the position in the public domain is far less clear with the sceptics receiving a disproportionate share of media attention. How does this confused view of science affect the public and which side do they believe? This chapter will look at some at the accusations levelled at scientists on both sides of the fence and the role scientists have played in politicizing global warming. The important question is, are we all prepared to reduce carbon dioxide emissions? The chapter will briefly consider levers that can be implemented to bring about reductions in greenhouse emissions. The responsibility for solutions to global warming rests in all our hands, not just scientists and politicians. We need to decide the price we will pay for our comfort and our environment.

Scientific certainty

The illusion of scientific certainty has probably slipped somewhat in recent years. There was a time when science was seen as a way of moving forward to some inexorable truth. That the pronouncements of scientists were absolute, born of objective deductive reasoning and little else. The work of Kuhn questioned this sterile environment, seeing science as a more social phenomenon. Kuhn (1962) in considering scientific controversies likens them to revolutions where a scientific paradigm is held to be true by the majority of scientists with only a few outsiders questioning the paradigm. The majority seek to maintain the paradigm and defend it from the outsiders' criticisms. However, eventually the cracks in the theory are too large to maintain and a new paradigm emerges to be supported by the majority. Although there is a gain in understanding, there is also a loss of some previously held beliefs. Scientific certainty is no longer about some objective truth but that which is accepted by a majority of scientists. It is easy to interpret the present situation as global warming being the current scientific paradigm and with the sceptics as the outsiders attempting to overthrow the standard view. The following section shows that it is global warming that has gained acceptance. Revolutions are usually absolute. In this way the new standard is set and also the new research problems to pursue are opened up and the old ones closed down. Even if there is no outright winner some sort of closure is required, some consensus of opinion, that allows science to move forward. Many sceptics would argue that this closure is taking place prematurely. The sceptics claim that research grants are denied them and journals refuse to publish their papers because they dare question the accepted global warming paradigm. Therefore many of their assertions appear on websites or non-academic journals rather than in peer-reviewed publications – for example, see Lindzen on the CATO website (Lindzen, 1999).

So where is the uncertainty in global warming? There is agreement that greenhouse gases are rising. Glacial cycles are a red herring occurring on much longer time-scales than global warming. The real scientific debate hinges on how sensitive the climate is to increasing greenhouse gases. Sceptics argue that the increasing greenhouse gases will make little difference to the climate but the majority of scientists beg to differ. So when will absolute certainty arrive? The 1990 IPCC report (Houghton *et al.*, 1990) makes clear that uncertainties will always remain and that trying to predict when scientists will be sure is probably more difficult that predicting future climate change.

the 'unprecedented change' theme to come up with some
. By assuming that the past 0.5°C warming is all due to nat-
and that another 0.5°C warming is needed to convince every-
g the worst environmental case (that there is no attempt to
ns and the climate is assumed to be highly sensitive) then it
il the year 2002 to detect change. However, if the lowest emis-
used with the least sensitive climate, then it would take until
2047 to detect the change.

The 1995 IPCC report (Houghton *et al.*, 1996) takes an entirely different approach stressing that the change will evolve gradually out of the climate signal. It says that global level changes will probably be detected before regional ones. However, the large uncertainties involved in both the signal and noise mean that the climate community is split. Some believe that as yet the uncertainties are too great to attribute any change to a human influence while others believe the signal has already been detected. Unambiguous attribution is as yet beyond science but a majority of scientists believe there is a 'discernible human influence on global climate' (Santer *et al.*, 1996). How science evolves, and the motives of scientists could be said to be purely academic if science does not influence the public. Scientific uncertainty merely leads to a fiercer scientific debate, it is hardly of concern to the person in the street. For example, the theory of continental drift and plate tectonics was subject to years of scientific argument but while interesting it did not affect people outside the scientific community (see Bridgstock *et al.* (1998) for an overview of the debate). Global warming, however, belongs to a different category of scientific theories. To counter it may require economic and social changes which require political decisions to be made. The theory of global warming became politicized.

The politicization of global warming

The work of Revelle and Suess (1957) really gave scientific credence to modern-day worries about global warming. In a review of the growing awareness of global warming Kellogg (1987) reports on the early non-scientific interest in the topic. In 1963 a USA **non-governmental organization** (NGO), the Conservation Foundation, organized a meeting on global warming. It concluded that a doubling of carbon dioxide in the atmosphere would lead to an increase of global temperatures of 3.8°C (Conservation Foundation, 1963). There followed in 1965 a publication by the President's Science Advisory Committee on water and air pollution. The report included the first consideration by the US government of global warming (Kellogg, 1987). The early 1960s was also the time when the new environmental movements began to be formed. Environmental concern can be traced back to the height of the industrial revolution. The nineteenth-century British movements, such as the National Trust, and the American Preservation of the Wilderness are facets of early environmentalism. They differ from the modern environmental movements that grew out of the 1960s in that they essentially strove to conserve nature with no political aims. The environmental groups of the 1960s, such as Greenpeace and Friends of the Earth, wanted direct political action, although they tended to act against the political system rather than with it (Pepper, 1984).

Concern for the environment was maintained on a wide range of issues through the early 1970s. The Study of Critical Environmental Problems report (SCEP, 1970) looked at many areas with a view to raising both public and scientific interest. This report considered one of the most pressing problems at the time which was the possible effect of supersonic transports, such as Concorde, on the stratospheric ozone layer. It also reviewed global warming. Several large-scale high profile scientific meetings followed. The United Nations conference in Stockholm in 1972 led to the founding of the United Nations Environment Programme (UNEP) (Kellogg, 1987). At this time there was considerable scientific doubt about global warming. It was dismissed as insignificant and cooling was a strong possibility (SCEP, 1970; Rasool and Schneider, 1971). Following the SCEP report an editorial in the journal *Nature* concluded that, 'half a degree (warming) by the end of the century ... is much more than sober men should worry about' (*Nature*, 1971). Ironically this is exactly the temperature change we are worrying about. By the mid-1970s public interest, had waned somewhat in all environmental issues and global warming took a back seat. There are several reasons why this happened. It has been noted that concern for the environment is greatest when the economy is booming and then declines at times of recession (Pepper, 1984). Lowe and Goyder (1983) note that environmentalism grew in Western democracies in the 1890s, 1920s, the late 1950s and early 1970s. These are all times of economic growth. The environmentalism of the early 1960s in Europe and North America also accompanied other fundamental societal changes and was a time for challenging authority and protest. By the mid-1970s the mood had begun to change as the economic down-turn took over. Another reason could be that issues are subject to an 'issue-**attention cycle**' (Downs, 1972; Portney 1992). First, issues are brought to the attention of the public in a dramatic way, focusing on the evils of the problem even though it may have been known about for a long time. Next there is public alarm and demands for something to be done, followed by a belief that the problem can be solved. Finally there is a decline in public interest as it is realized that the problem is far more difficult and expensive to solve than originally thought. Other issues then arise which require the public's attention. There certainly were other issues to grab the public attention. There arose a series of debates, and important counter-debates, about the potential effect of humans on the environment. In the early 1970s the oil crisis broke. Fuel was in short supply and prices rose. This provided ample material for reports such as *Limits to Growth* (Meadows and Meadows, 1972). This was followed by the revelation by Molina and Rowland (1974) that CFCs could deplete the ozone layer. The debate about CFC theory was particularly intense in the USA and several states brought in legislation banning CFCs in aerosols. Later in the early 1980s came studies into the effects that a nuclear war might have on the climate. Climate models predicted that the soot and other debris thrown into atmosphere would sufficiently increase the **atmospheric turbidity** to block out the sun. Global cooling would occur and a **nuclear winter** would follow which would lead to widespread failure of any agricultural crops planted by those lucky enough to escape the nuclear conflagration (Cess, 1985; Covey *et al.*, 1984; Teller, 1984). Then came the discovery of the 'ozone hole' over Antarctica and the subsequent focus on international legislation to tackle stratospheric ozone depletion (Farman *et al.*, 1985).

While public interest in global warming waned, scientific interest did not. As Kellogg (1987) relates, there continued to be large-scale international conferences addressing the issue throughout the 1970s and early 1980s. Many of these conferences were sponsored or hosted by US scientists. Governmental involvement at this stage was limited to interested parties such as the USA Environmental Protection Agency (EPA). Mazur and Lee (1993) contend that the low interest shown by the US government in the mid-1980s was partly due to the other environmental problems and the recovery from the 1970s' oil crisis. Cheap, plentiful fuel removed the impetus to find alternatives that might be more environmentally friendly. They also suggest that scientists' interest was diverted by the nuclear winter and stratospheric ozone debates. However, the evidence for this is not conclusive and global warming was able to command centre stage at important scientific meetings, first, at Villach, Austria in October 1985 and then two workshop meetings in 1987. The first of these in September was also in Villach and the second in November was in Bellagio, Italy. Not only was the global warming trend 'confirmed', but also policies designed to address reducing greenhouse gas emissions were discussed for the first time (Rowlands, 1992). Global warming was accepted not just by rank and file scientists but also 'blue ribbon meteorologists' (Kellogg, 1987). Thus the elite scientists, those scientists who interacted closely with government, who had responsibility for the direction of large-scale scientific programmes, could fully endorse the theory.

The global warming debate may have moved on but it was still within the civil service of government: those who serve politicians. The turning point, the politicization of global warming, is often identified as occurring in June 1988. The USA was experiencing one of the worse droughts on record which was seriously affecting agricultural production. The *New York Times* reported on 17 June 1988 that stored grain was the lowest for 5 years and would last for only 77 days (*New York Times*, 1988). A congressional hearing into the matter was headed by Senator Timothy Wirth. On 23 June 1988, a day which was the hottest day of the year in Washington DC, the congressional meeting listened to a statement from James Hansen, then chief climate scientist for NASA. The day turned out to be one of the hottest on record and Hansen declared that 'the greenhouse effect is here' (Hare, 1988). Hansen directly linked the drought and the warm weather to global warming. The congressional hearing had been well attended by the press and so media coverage was high. At the same time the Toronto conference was being held. This was a meeting of scientists but the US drought combined with Hansen's statement ensured it was well covered by the media (Jäger and O'Riordan, 1996). The conference endorsed Hansen's statement and concluded that the time had now come to act. It called for developed countries to reduce their greenhouse gas emissions by 20 per cent from 1988 levels by the year 2005. Global warming was now firmly on the political agenda (Paterson, 1996; Rowlands, 1995). The Toronto conference paved the way for widespread political acceptance and led to the establishment of the IPCC in December 1988 by the UN.

The role of scientists

In the foregoing section the motives of scientists have barely been questioned. What role did scientists play in politicizing global warming and should that affect our belief in the message? Kuhn (1962) was amongst the first of many to question the traditional view of the scientific method. The motives of scientists have also been questioned. There is a view that most pure science is pursued out of curiosity; that scientists give little or no thought to the practicalities or outcomes of their research (Polanyi, 1962). Thus science is seen as providing neutral, value-free evidence to policy-makers. This idea is widely criticized but still thought by some to be the generally accepted model of science by the great majority of people (Lipschutz and Conca, 1993). A difficulty for scientists is trying to promote the view that global warming is a problem to be addressed and at the same time remaining honest to the scientific tradition (Schneider, 1988). The extent to which funding and self-aggrandizement influence the motives and views of scientists is as fierce as the debate about global warming. There are those who would argue that scientists are sufficiently bound to a scientific code of honour that the source of their funding has little effect on their findings (Bridgstock *et al.*, 1998). Others would beg to differ. In particular the scientific sceptics have been heavily criticized for accepting funding from organizations campaigning against limits to greenhouse gas emissions (Ozone Action, 1996). By implication, the source of their funding has influenced their results. The sceptics have hit back by pointing out that funding from government sources to those scientists supporting the greenhouse theory far exceeds anything they receive.

Haas (1992) developed the idea of epistemic communities promoting policy action. This means groups of scientists reaching a scientific consensus on an environmental issue and then promoting that issue to achieve a political change. The idea was first developed to explain the progression of stratospheric ozone depletion from scientific debate to international legislation. In global warming the application is far from clear. Some argue that the first step was achieved with a broad scientific consensus being reached but that since the issue reached the political agenda scientists have been shut out. Boehmer-Christiansen, a respected social scientist, uses the idea of epistemic groups to argue that the IPCC have acted to make their services invaluable to policy-makers, thus ensuring a steady stream of funding. Boehmer-Christiansen (1994a, b) contends that elite scientists from a variety of national institutes, who went on to establish the IPCC, first, over-emphasized the threat from global warming thus making their research relevant to policy decisions. Second, that by achieving a consensus view the IPCC were able to consolidate their position. Finally, once global warming was the subject of international negotiations the IPCC emphasized the uncertainty of global warming, thereby making sure of continued research funds and relevance to national and international policies. This final point also suits the politicians, as the focus on uncertainty means that governments can implement or delay certain policies as it suits them. Boehmer-Christiansen (1994b) questions the role that science can play in policy decision-making, particularly as the idea of science as a positive rationale arbitrator is increasingly rejected. She considers that government should play a stronger role seeking environmental advice from many different quarters. Jäger and O'Riordan (1996) argue that physical scientists have little more to offer the global warming debate and that real effort should be expended on multi-disciplinary teams, including both physical and social scientists, addressing the impacts of global warming. Therein, however, lies a fundamental divide which is illustrated by the views offered in the previous paragraph. Many elite scientists, apparently, still believe in the essential purity of science, proffering knowledge without judgement. This is a position that most social scientists, whatever their theoretical perspective, find hard to accept. Jäger and O'Riordan (1996) point out that it is almost futile for scientists to expect their pronouncements not to have some sort of political ramifications.

Science no longer stands on the sidelines but now plays a strong role in setting the agenda of environmental priorities and defining environmental risks. Scientists are seen as not offering advice but pronouncements, not sharing knowledge but dictating it. Thus scientists have become part of the political framework and power structure of government. It is claimed that scientists and their science have become 'corrupted' by this close involvement with policy decisions (Lipschutz and Conca, 1993). Others perceive this intimate relationship between science and policy as threatening any real environmental management and that the scientific tradition may lead to increased environmental degradation rather than improvement (O'Riordan, 1995). These increasing attacks on scientific credibility have been interpreted by others, notably those in the scientific community, as a lack of understanding of science by the public. If only the public were better educated as to scientific method they would understand its benefits more and be less irrational in their response to risks in the environment (Portney, 1992). This, coupled with the Western belief that economic success goes hand in hand with scientific and technological improvements, means that there are constant calls to increase the public understanding of science (Dunbar, 1995; Gillott and Kumar, 1995; Royal Society, 1985). Underlying these sentiments is what Portney (1992) terms a clash between science and humanism. Is it science that determines the direction of society, or should society determine the direction and only use science where appropriate? Portney (1992) relates how this clash is common in environmental issues, particularly when considering the risk to society of environmentally impacting technologies.

There are currently two predominant theories which deal with science, society and the environment. These are the **risk society theory** (RST) and **ecological modernization theory** (EMT). Both theories come from German schools of thought on global environmental change and reflect opinions from opposite ends of the spectrum. On the far green side are those who think that in order to become harmonious with the environment our whole way of life has to change and the structure of society radically altered. At the other end are those that see science and technology as the very solution to all the problems created by science and technology. Both these theories are based on Western constructs and ignore links between wealth and the environment. Many of the ethical issues raised relate to the north–south divide, to the developing and developed world (Blowers, 1997). Therefore these theories cannot said to be universal but probably apply only to the industrialized north.

RST was first put forward by Beck (Beck, 1992; Giddons 1990) and is part of a much wider philosophical debate about how society operates as the manufacturing age comes to a close. Taking a narrow environmental viewpoint, RST offers a rather pessimistic outlook. Modern society has become one of ecological catastrophically disastrous risks. The risks are self-afflicted arising from the modern way of life. In particular it refers to the nuclear age. Nuclear power benefits us all but in creating it we have also created a risk that power plants might explode destroying all life. Such a way of life also puts us at the mercy of experts, scientific experts who control the technology and try to persuade the public of its acceptability. Individuals then feel powerless against the experts and unable to change the outcome. The only way to avoid the risk is to not to partake in them. This would seem to lead to an inevitable global environmental catastrophe unless somehow self-awareness allows a transformation of attitudes and values. In contrast, EMT, originally proposed by Huber (Hajer, 1996), is far more optimistic. EMT perceives science and technology as a way to achieve sustainable development. Industry, through free trade, should be encouraged to see the environment as an opportunity for growth, using clean technology, recycling, and by changing consumption and production. It is in a sense business as normal but with due regard to the environment. It fits well with current Western political rhetoric and does not challenge the status quo or question the ethical issues raised (Blowers, 1997). Most importantly EMT proposes that economic growth is reconcilable with care for the environment (Hajer, 1996).

Applying the solutions

In the initial euphoria after the signing of the 1987 Montreal Protocol protecting the stratospheric ozone layer it was thought that this was a model piece of legislation. This could show the way to tackling other environmental problems, most notably global warming (Brenton, 1994). It was, after all, a piece of legislation showing international co-operation based on uncertain science that forced industry, and the rest of us, to change. Since then, however, a different view has emerged. Although the Montreal Protocol is predicted to allow the ozone layer to heal by 2050, this will only happen if the developing countries and black-market operations cease production of CFCs. DuPont, one of the leading suppliers of CFCs had patents relating to replacement chemicals dating from the 1970s. It has been argued that the response and initial delaying tactics of the main producers of CFCs was driven primarily out of economic consideration (Howes *et al.*, 1997). Furthermore, this was essentially a technical fix requiring no real change in society's operations. People are still able to buy fridges and aerosol cans. Global warming will require much more widespread changes and legislation if cuts in emissions are to be made (Greene and Skea, 1997; Parry *et al.*, 1998).

The policy-making process is described as a cycle (*see* Fig. 8.1) and the broad policy options of mitigation and adaptation were discussed in Chapter 7. The different profiles of each country will, to some extent determine the nature of the policies that they pursue to reduce carbon dioxide emissions. For example, Table 8.1 shows the breakdown of EU carbon dioxide emissions by sector and power stations are revealed as one of the largest generators of carbon dioxide emissions. This figure, however, hides wide disparities between individual countries in the EU. In the energy sector some countries such as France and Sweden are heavily dependent on nuclear power, meaning that carbon dioxide emissions from power stations are low. In contrast, those countries with fossil fuel reserves, such as Denmark, Germany and Britain, have favoured these and so have higher carbon dioxide emissions (Collier and

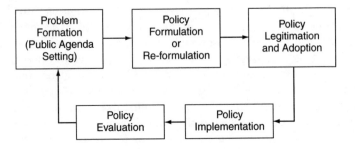

Figure 8.1 The policy-making process in the USA
Source: Portney (1992)

Table 8.1 Percentages of carbon dioxide emissions by sector for 1996

	World	USA	EU	UK
Electricity and heat production	32.1	35.8	25.9	28.4
Industry	21.6	12.1	17.3	14.8
Transport	20.2	30.5	26.0	25.1
Other unallocated energy industry and autoprocessors	10.7	9.3	9.5	10.4
Other (of which residential)	15.4	12.3 (7.2)	21.3 (13.3)	21.3 (14.9)

Source: IEA (1998) raw data in GtC/yr

Löfstedt, 1997a). Therefore this section will only review some of the more prominent policy levers that might be used to encourage acceptance of environmental policies. These levers could be broadly categorized as voluntary agreements and state intervention.

Industry associations often promote free market trade and voluntary agreements as the best way forward for industry to tackle environmental issues including global warming. In Britain this was the main strategy of the government throughout most of the 1990s (Howes *et al.*, 1997). However, the predicted savings from these energy efficiency measures were constantly revised downwards. Other countries, such as Germany, have taken a route dominated by legislation and have tried to give a lead on environmental policies (Huber, 1997). Legislation has advantages for industry because it sets a level playing field and thus avoids the dilemma of choosing between good business practice and being green (Purvis *et al.*, 1997). Sustainable development holds up the promise to business that compliance with environmental legislation is not a cost but an opportunity for business. This has led to claims that in responding to the environment a business will profit: the so-called 'win-win' or **double dividend**. One of the leading proponents of this is Michael Porter (1991). He has developed a hypothesis that increased environmental legislation will cause businesses to be innovative and develop new processes. In doing so they will increase their competitiveness because the innovations will lead to increased quality and lowered cost of production. This will more than offset any costs due to addressing the environmental regulation. However, merely complying with legislation will not be enough. Those businesses that are proactive about the environment and seek standards higher than the legislation, called compliance-plus, will be the ones to reap the rewards (Howes *et al.*, 1997). Businesses that would be able to make the most of the environmental opportunity would be those that took on board the environment, making it part of their business structure from management to strategy. This led to the idea that businesses could be categorized as to the extent of their 'greenness' (Welford and Gouldson, 1993). Although it is debatable whether it was the original intention, these categorizations came to be seen as a development cycle through which businesses would go in order to achieve full environmental integration. It seems more likely, however, that rather than follow a progression businesses will pick and choose the strategies that are right for them (Hass, 1996).

The evidence for both the double dividend and the greening of business is distinctly patchy. There are some impressive examples of win-win cases where businesses have made significant gains by addressing an environmental problem. In Britain these are most often associated with waste minimization programmes. Impressive gains were made in the USA through reductions in air pollution. While such outcomes exist, however, they would appear to be rare (Walley and Whitehead, 1994; Howes *et al.*, 1997). The win-win dividend also proposes that the manufacturing process is completely changed to one of clean technology. An end-of-pipe solution is seen as inefficient. In reality the distinction between the two is rare. Howes *et al.* (1997) conclude that most businesses rarely make investment decisions on environmental grounds alone. They say that increased environmental legislation will lead to winners and losers. Most research devoted to the 'greening of business' and the benefits to business of environmental awareness has concentrated on large corporations. Only recently has attention begun to be paid to small and medium sized enterprises (SMEs). SMEs are important; for example in Britain they employ up to half the workforce. The perception tends to be that multinationals have the resources and willingness to respond to environmental issues whereas SMEs do not have the capability or the motivation to do so. Such sweeping generalizations hide some anomalies. Some SMEs are set up principally as 'green' businesses reflecting the environmental ethics of the owner-manager (Drake *et al.*, 1997). Large corporations are much more susceptible to a fragmentation of greening; with some sectors of a company being more environmentally aware than others. Also, while multinationals may well have formalized their approach to the environment, they have also learnt how to manage the environmental policy process. Thus industry lobbies for change to environmental policies and slows down their development (Howes *et al.*, 1997).

With regard to global warming, Howes *et al.* (1997) note that while some board decisions may have been taken with reference to future legislation, on the whole 'industry seems largely unaffected by the issue'. The industry response to global warming appears fragmented. American industry is far more antagonistic towards emission reductions than European industry. This may reflect greater public awareness or concern within Europe over global warming (Howes *et al.*, 1997). For instance, one might expect that the large oil-producing corporations would naturally oppose any legislation. However, even here there are differences between America and Europe. The USA oil-producers have been very active against any legislation to reduce greenhouse gas emissions. The European-owned companies have tended to take a more neutral stance and have started to widen their operational base (ENDS, 1999). The main business sector supporting legislation to address global warming is the insurance industry. If global warming does lead to increased incidence of severe weather events, then the insurance industry will be the ones paying out (Spencer-Cooke, 1999). Non-governmental organizations tend to be associated with the green lobby but there are also plenty arguing against legislation that interferes with the free market. These often take the form of non-profit-making right-wing think-tanks (Beder, 1997; Beder, 1999). Thus some of the most vociferous claims against global warming come not from natural scientists but social scientists and economists supporting the *laissez-faire* approach (for example see Morris, 1997). They are far more common in the USA than in Europe; as a search of the World-Wide Web will confirm (see Appendix for Websites). One of the most prominent is the Global Climate Coalition, supporting the views of the US fossil fuel industries and other like-minded industries and trade associations (Retallack, 1999). Their agenda is to resist any agreements that may reduce US competitiveness (Beder, 1997).

While industry and business are often the targets of high profile campaigns by environmental groups and are often seen as the bad guys needing to clean up their act, they are only part of the problem. In the EU industry produces around 25 per cent of carbon dioxide emissions and this has been falling due to changes in the industrial make-up of business, with the move from a manufacturing base to a service base (Collier and Löfstedt, 1997a). Table 8.1 reveals that, in the EU, other, non-industrial, sources comprise about 20 per cent of all carbon dioxide emissions. Perhaps as much as 15 per cent of carbon dioxide emissions are due to residential activity. We all consume energy, we all use transport, we are all part of the problem. It is not just industry and power generators that need to change their ways. To tackle global warming requires us all to reduce our consumption. How are we, the public, to be encouraged to reduce our consumption? At one point it was hoped that green consumerism, the desire for the public to buy environmentally sound products, would be enough. This now seems an unlikely route to achieve the emission savings required. Most of the policy levers that can be applied to industry can also be used on the individual. Legislation can be introduced to increase the efficiency of domestic goods. By enforcing the production of more efficient white goods such as fridges and washing machines, consumers are obliged to buy one – and to pay for it. Although, as noted in the previous chapter, this may not have the desired effect and may result in increased energy consumption. Grants, subsidies and investments can all be made to work for lower emissions. For instance, investment can be made in public transport, with subsidies for ticket prices, to encourage its use in preference to the private car. A major challenge facing politicians is to introduce these changes while maintaining economic growth and staying in office. Perhaps one of the most controversial policy levers is a carbon tax. This illustrates the dilemma facing politicians, as voters are not well known for voting for higher taxes (Dickson, 1993). The EU originally wanted to implement European-wide carbon taxes but national differences, particularly over sovereignty thwarted early attempts. While Britain was particularly vocal in its opposition, other countries were also happy for the tax not be implemented. Therefore by 1997 only Austria, Denmark, Finland, the Netherlands and

Norway had implemented a carbon tax. These have predominantly concentrated on the domestic sector leaving industry untouched. Even Britain imposed VAT on domestic fuel (Collier, 1997; Dickson 1993). However, these taxes have been raised at a time when energy prices are being forced downwards and it is unclear whether they will have any effect on overall consumption (Collier, 1997). The Kyoto protocol has led to new changes. In Britain a tax has been introduced on the energy used by industry (Pearce, 1999a).

Will these changes be enough? Collier and Löfstedt (1997b) conclude, 'If reasonably stable unions of people such as the EU cannot find ways of addressing climate change issues it is difficult to see how it can be addressed at all.' The levers introduced in this section represent what might be termed a scientific response to the problem of reducing energy consumption. Shove *et al.* (1998) boil this policy down to two questions: 'why won't people save energy?' and 'what will influence them to do so?'. These policy initiatives assume that energy efficiency is about pushing the right levers. Shove *et al.* (1998) argue that people cannot be categorized by their energy usage. They say that people's attitude to energy usage is socially constructed; our habits are culturally, historically and socially defined. This leads to the conclusion that current policy initiatives are doomed to failure, except in certain limited situations. Shove *et al.* (1998) call for a much wider approach to energy policy which encompasses social processes in order to find new initiatives.

And finally . . .

In the European Union only two countries, Britain and Germany, are likely to return to 1990 levels of greenhouse gas emissions by the year 2000, the target agreed at Rio in 1992. These are widely acknowledged to be due to economic factors, however, not due to environmental pressure. In Britain it is due to recession and the 'dash for gas', the conversion from coal to gas-fired power stations. In Germany it is the effects of reunification with eastern Germany (Collier, 1997; Huber, 1997; *Nature*, 1997). By mid-March 1999 80 countries had signed the Kyoto Protocol. Some countries like the Maldives are motivated by the need to save their country from rising sea levels. However, others have failed to sign. Iceland, which is allowed to increase its emissions by 10 per cent, has refused to sign. The Icelandic government wants the emissions level increased to 25 per cent to accommodate a new aluminium smelter (Pearce, 1999a). Most importantly, however, the United States of America at the forefront of the early environmental negotiations, such as the Montreal Protocol, has increasingly taken a back seat on international environmental agreements. At Kyoto in December 1997 an accord on greenhouse gases was reached with the agreement of the US. While the 5.2 per cent reduction in greenhouse gas emissions was seen as a victory for the EU, the US gained a victory in the flexible trading of emissions. The emissions of the six greenhouse gases, carbon dioxide, nitrous oxide, methane, HFCs, perfluorocarbons (PFCs) and sulphur hexafluoride (SF_6), can be traded off between each other. Thus higher emissions of one gas can be offset by lowering emissions of another. The emissions can also be traded between countries allowing some countries to sell off their surplus (ENDS, 1997). By summer 1998 the US Congress voted not to accept the cuts in greenhouse gas emissions unless developing countries did the same, despite the inequity in emissions. Indeed, there are claims that Congressional moves will outlaw the US government and its agencies from even debating global warming or researching into it (Pearce, 1998). The US signed the treaty at the CoP-4 meeting in Buenos Aires, but continues to negotiate changes in the emission targets of certain developing countries before cutting its own emissions (Pearce, 1999b). It has been suggested that without implementing policies to cut its greenhouse gas emissions by the year 2010 US emissions will be 34 per cent above 1990 levels. In September 1999 Dr Klaus Töpfer, head of UNEP, announced that it was 'too late to stop global warming' and has been interpreted as a warning to nations that without further commitment the Kyoto Protocol will fail (McCarthy, 1999).

It would appear that politicians are sceptical of the public's commitment and the ability of sustainable development to produce sustained growth, whatever the rhetoric. This is despite numerous opinion polls all suggesting that we, the public, have a growing concern about the environment (Portney, 1992). The level of concern over global warming exhibited in polls varies widely between countries (Hofrichter and Klein, 1991; Rudig, 1995). The polls also reveal that the public appeared confused over exactly what global warming is and how it should be tackled (Rudig, 1995; Bord et al., 1998). However, in-depth studies of people's attitudes show that greater understanding about global warming may not necessarily lead to people changing their behaviour (Staats et al., 1996). Polls conducted in the US prior to the Kyoto Protocol suggested that the public reject the sceptics' view and accept that global warming is happening. Three quarters of those interviewed thought that the only scientists who did not believe in global warming were in the pay of the oil, coal and gas companies (Macilwain, 1997a). However, it is unclear whether expressing concern in a poll translates into actually voting for a change which may involve extra expense and inconvenience (Portney, 1992). It has been suggested in the US that environmental concerns can significantly affect voting. In the early 1970s there were widespread concerns over the effect of commercial supersonic transports (SSTs) on the stratospheric ozone layer. It is claimed that senators voted for the cancellation of the US SST in 1971 because they feared that otherwise the public would vote them out of office in the upcoming elections (Rosenbloom, 1981). Mitchell (1984) also claims that voters are willing to pay for the change. However, the evidence is not clear-cut. In Britain, when asked, over half of people say they would buy environmentally friendly goods and yet sales of items such as organic food remain low (ESRC, 1998). Opinion polls show people to be ambivalent to taxes, wanting any carbon tax to be offset by reductions in other taxes. People are concerned but they do not want their lifestyle to be compromised (Bord *et al.*, 1998; Staats *et al.*, 1996). The public also appear sceptical that any one policy can address global warming (Rudig, 1995). As noted earlier in this chapter, the state of the economy also seems to shape people's opinion of the environment and this is reflected in opinion polls (Portney, 1992). People's perception of the problem also affects their attitude. Global warming is often seen as too large an issue for the individual to respond to and therefore nothing can be done. Also, in colder climates, such as Scotland, global warming can be perceived as a good thing (ESRC, 1998).

The media also have a strong role in shaping the public's perception. The sceptics, though they are in the minority, have done well in getting their message across (Beder, 1999). This may have as much to do with political correctness and the need to appear even-handed on the part of the press as it does the sceptics' ability to hold press conferences. It does mean, however, that even at the end of the 1990s articles appear in newspapers declaring that global warming will herald a new 'Garden of Eden' (Avery, 1999). More powerfully in *The Wall Street Journal* in June 1996 there appeared an open letter, by a former president of the US National Academy of Science, attacking the 1995 IPCC report and the lead authors of WG-I (Glantz, 1997). Perhaps it is not surprising that such an article should be printed in a financial newspaper. It is in the economics that the sceptics would appear to have their greatest weapon. It plays on fears that cuts in greenhouse gas emissions will lead to depressed economic growth (Macilwain, 1997b; Retallack, 1999). There are those in the developed world who argue that given the uncertainty of the science coupled with the economic risk there should be a 'wait and see' policy. It is also argued that draconian environmental legislation limiting greenhouse gases will hold back the developing nations. Some developing nations have already suggested that for them a preferable route is to adapt to climate change fearing limits on their ability to develop. For many developing nations greenhouse gas emissions per capita remain very low and in the list of problems that these countries face, global warming ranks very low (Rahman *et al.*, 1993). Others call for a **'no regrets' policy**, whereby only those measures that have clear social benefits should be implemented (Clayton, 1995; Howarth and Monahan, 1996). In contrast, the environmental lobby argue for a strong sustainability approach. To some extent opinion polls would suggest that the public support this position. American opinion polls show that the public believe that the environment should and can be protected. However, the public want this to be done without affecting economic growth (Portney, 1992; Walley and Whitehead, 1994). It is the economy, not the environment, that the public identify as the most important issue affecting the country (Portney 1992).

Perhaps the greatest hope for changing attitudes to the environment is through education and evolving social consciousness. Much like the legislation that has grown to protect children and promote their welfare, so environmental legislation will grow. Environmental concern will become encompassed in all our daily practices so much so that they will be taken for granted. Such societal changes, however, tend to come gradually, taking generations to achieve. Global warming requires action soon. The understanding of global warming begins with science and the understanding of climatic change. The implications of global warming and the involvement of science, however, extend far beyond the traditional boundaries. The reason why global warming remains such a contentious issue is because it is so much more than a scientific theory. The science of global warming is complex and challenging but ultimately an environmental issue on such a vast scale as this challenges far more than our knowledge and understanding. It calls into question our values of science and economics versus humanism. It questions the role of nature and government in our society (Portney, 1992). At the beginning of the new millennium there are some tough decisions to take. Research into global warming and climate change will provide an increasingly rich backdrop of scientific information. While the scientific consensus surrounding global warming may grow, the difficulties of confronting this problem and finding policy solutions become all the more pertinent. Scientific 'facts'can be interpreted in a wide variety of ways to fit a range of prejudices and desires. It would be foolish to think that science alone can provide the answer. Global warming requires a much broader response. It requires policy-makers, and all of us, to take decisions that affect our world's welfare.

References

Avery, D.T. 1999: Welcome to the Garden of Eden. Saturday Review *The Guardian* 15 May.

Beck, U. 1992: *Risk society: towards a new modernity.* London: Sage.

Beder, S. 1997: *Global spin: the corporate assault on environmentalism.* Totnes: Green Books White River Junction and Chelsea Green Publishing Company.

Beder, S. 1999: Corporate hijacking of the greenhouse debate. *The Ecologist* 29, 119–22.

Blowers, A. 1997: Environmental policy: ecological modernisation or the risk society? *Urban Studies* 34, 845–71.

Boehmer-Christiansen, S. 1994a: Global climate protection policy: the limits of scientific advice – part 1. *Global Environmental Change* 4, 140–59.

Boehmer-Christiansen, S. 1994b: Global climate protection policy: the limits of scientific advice – part 2. *Global Environmental Change* 4, 185–200.

Bord, R.J., Fisher, A. and O'Connor, R.E. 1998: Public perspectives of global warming: United States and international perspectives. *Climate Research* 11, 75–84.

Brenton, T. 1994: *The greening of Machiavelli: The evolution of international environmental politics.* London: Earthscan.

Bridgstock, M., Burch, D., Forge, J., Laurent, J. and Lowe, I. 1998: *Science and technology and society: an introduction.* Cambridge: Cambridge University Press.

Cess, R.D. 1985: 'Nuclear war': illustrative effects of atmospheric smoke and dust on solar radiation. *Climatic Change* 7, 238–51.

Clayton, K. 1995: The threat of global warming. In O'Riordan T. (ed.), *Environmental science for environmental management.* Harlow: Longman Group.

Collier, U. 1997: 'Windfall' emission reductions in the UK. In Collier, U. and Löfstedt, R. E. (eds), *Cases in climate change policy: political reality in the European Union.* London: Earthscan.

Collier, U. and Löfstedt, R. E. 1997a: The climate change challenge. In Collier, U. and Löfstedt, R. E. (eds), *Cases in climate change policy: political reality in the European Union.* London: Earthscan.

Collier, U. and Löfstedt, R. E. 1997b: Comparative analysis and conclusions. In Collier, U. and Löfstedt, R. E. (eds), *Cases in climate change policy: political reality in the European Union.* London: Earthscan.

Conservation Foundation 1963: *Implications of rising carbon dioxide content of the atmosphere.* New York: The Conservation Foundation.

Covey, C., Schneider, S.H. and Thompson, S.L. 1984: Global atmospheric effects of massive smoke ejections from a nuclear war: results from general circulation model simulations. *Nature* 308, 21–5.

Dickson, D. 1993: British voters force rethink of global warming strategy. *Nature* 364, 372.

Downs, A. 1972: Up and down with ecology: the 'issue-attention cycle'. *The Public Interest* 2, 38–50.

Drake, F., Hunt, J. and Purvis, M. 1997: Small companies, big responsibilities. In Greene, O. and Skea, J. (eds), *ESRC Global Environmental Change, Special Briefing 1.* Brighton: ESRC Global Environmental Change Programme.

Dunbar, R. 1995: *The trouble with science.* London: Faber and Faber.

ENDS 1997: The unfinished climate business after Kyoto. *ENDS report no. 275,* 16–20.

ENDS 1999: BP Amoco: learning to live with a CO_2 ceiling. *ENDS report no. 294,* 19–22.

ESRC 1998: *A climate change mitigation strategy for Scotland, Workshop report.* Edinburgh: Economic and Social Science Research Council and the Scottish Office.

Farman, J.C., Gardiner, B.G. and Shanklin, J.D. 1985: Large losses of total ozone in Antarctica reveal seasonal ClO_x/NO_x interaction. *Nature* 315, 207–10.

Giddens, A. 1990: *The consequences of modernity.* Cambridge: Polity Press.

Gillott, J. and Kumar, M. 1995: *Science and the retreat from reason.* London: Merlin Press.

Glantz, M.H. 1997: Media wars over climate change. *Network Newsletter* 12 (no. 2), 1.

Greene, O. and Skea, J. (eds) 1997: After Kyoto: making climate change policy work. *ESRC Global Environmental Change, Special Briefing 1.* Brighton: ESRC Global Environmental Change Programme.

Haas, P.M. 1992: Banning cholorofluorocarbons: epistemic community efforts to protect stratospheric ozone. *International Organizations* **46**, 187–224.

Hajer, M.A. 1996: Ecological modernisation as cultural politics. In Lash, S., Szerszynski, B. and Wynne, B. (eds) *Risk, environment and modernity: towards a new ecology.* London: Sage.

Hare, K.F. 1988: Jumping the greenhouse gun? *Nature* **334**, 646.

Hass, J. 1996: Environmental ('green') management typologies: An evaluation, operationalization and empirical development. *Business Strategy and the Environment* **5**, 59–68.

Hofrichter, J. and Klein, M. 1991: *Evolution of environmental attitudes towards environmental issues 1974-1991.* Brussels: DG XI Environmental Unit.

Houghton, J.T., Jenkins, G.J. and Ephraums, J.J. (eds) 1990: *Climate change: the IPCC scientific assessment.* Cambridge: Cambridge University Press.

Houghton, J.T., Meira Filho, L.G., Callander, B.A., Harris, N., Kattenberg, A. and Maskell, K. (eds) 1996: *Climate change 1995: the science of climate change.* Cambridge: Cambridge University Press.

Howarth, R.B. and Monahan, P.A. 1996: Economics, ethics and climate policy: framing the debate. *Global and Planetary Change* **11**, 187–99.

Howes, R., Skea, J. and Whelan, B. 1997: *Clean and competitive: motivating environmental performance in industry.* London: Earthscan.

Huber, M. 1997: Leadership and unification: climate change policies in Germany. In Collier, U. and Löfstedt, R. E. (eds), *Cases in climate change policy: political reality in the European Union.* London: Earthscan.

IEA 1998: CO_2 *emissions from fossil fuel combustion 1971–1996.* International Energy Agency. Paris: OECD/IEA.

Jäger, J. and O'Riordan, T. 1996: The history of climate change science and politics. In O'Riordan, T. and Jäger, J. (eds), *Politics of climate change: a European perspective.* London: Routledge.

Kellogg, W.M. 1987: Mankind's impact on climate: an evolution of awareness. *Climatic Change* **10**, 113–36.

Kuhn, T. 1962: *The structure of scientific revolutions.* Chicago: The University of Chicago Press.

Lindzen R.S. 1999: Global warming: the origin and nature of alleged scientific consensus, *Regulation – The Cato Review of Business and Government.* http://www.cato.org/pubs/regulation/regl5n2g.html

Lipschutz, R. and Conca, K. 1993: *The state and social power in global environmental politics.* New York: Columbia University Press.

Lowe P. and Goyder, J. 1983: *Environmental groups in politics.* London: Allen and Unwin.

Macilwain, C. 1997a: Public rejects sceptics' line on global warming. *Nature* **389**, 531.

Macilwain, C. 1997b: US target 'likely to equal 1990 emissions'. *Nature* **389**, 531.

Mazur, A. and Lee, J. 1993: Sounding the global alarm: environmental issues in the US national news. *Social Studies of Science* **23**, 681–720.

McCarthy, M. 1999: 'Too late to stop global warming'. *The Independent*, 16 September.

Meadows, D.H. and Meadows, D.L. 1972: *The limits to growth: a report to the Club of Rome's project on the predicament of mankind*. New York: Universe Books.

Mitchell, R.C. 1984: Public opinion and environmental politics in the 1970s and 1980s. In Vig, N.J. and Kraft, M.E. (eds), *Environmental policy in the 1980s: Reagan's new agenda*. Washington DC: Congressional Quarterly Press.

Molina, M.J. and Rowland, F.S. 1974: Stratospheric sink for chlorofluoromethanes: chlorine atom-catalysed destruction of ozone. *Nature* **249**, 810–12.

Morris, J. 1997: *Climate change – challenging the conventional wisdom*. IEA Studies on the Environment, no. 10, London: IEA Environmental Unit.

Nature (editorial) 1971: The great greenhouse scare. *Nature* **229**, 514.

Nature (editorial) 1997: Seizing global warming as an opportunity. *Nature* **387**, 637.

New York Times 1988: Drought brings specter of major price increases. *New York Times* 17 June, 1.

O'Riordan, T. 1995: Environmental science on the move. In O'Riordan, T. (ed.) *Environmental science for environmental management*. Harlow: Longman Group.

Ozone Action 1996: Scientific skeptics on attack at international meeting on climate change. *Ozone action media release*, 10 July.

Parry, M., Arnell, N., Hulme, M., Nicholls, R. and Livermore, M. 1998: Adapting to the inevitable. *Nature* **395**, 741.

Paterson, M. 1996: *Global warming and global politics*. London: Routledge.

Pearce, F. 1998: Warring over warming. *New Scientist* **159** (no. 2144), 22.

Pearce, F. 1999a: Tax is on target. *New Scientist* **161** (no. 2178), 17.

Pearce, F. 1999b: A carbon fix? *New Scientist* **162** (no.2190), 22.

Pepper, D. 1984: *The roots of modern environmentalism*. London: Routledge.

Polanyi, M. 1962: The republic of science: its political and economic theory. *Minerva* **1**, 54–73.

Porter, M. 1991: America's green strategy. *Scientific American*, April, 96.

Portney, K.E. 1992: *Controversial issues in environmental policy: science vs. economics vs. politics*. Newbury Park: Sage.

Purvis, M.F., Drake, D., Clarke, D., Phillips, and Kashti, A. 1997: Fragmenting uncertainties: some British business responses to stratospheric ozone depletion. *Global Environmental Change* **7**, 93–111.

Rahman, A., Robins, N. and Ronceral, A. (eds) 1993: *Exploding the population myth: consumption versus population. Which is the climate bomb?* Brussels: Climate Network Europe.

Rasool, S.I. and S.H. Schneider 1971: Atmospheric carbon dioxide and aerosols: effects of large increases on global climate. *Science* **173**, 138–41.

Retallack, S. 1999: An interview with the Global Climate Coalition. *The Ecologist* **29**, 123–4.

Revelle, R. and Suess, H.E. 1957: Carbon dioxide exchange between atmosphere and ocean and the question of an increase of atmospheric CO_2 during the past decades. *Tellus* 9, 18–27.

Rosenbloom, J. 1981: The politics of the American SST programme: origin, opposition and termination. *Social Studies of Science* 11, 403–23.

Rowlands, I.H. 1992: International politics of global environmental change. In Rowlands, I.H. and Greene, M. (eds), *Global environmental change and international relations*. London: MacMillan.

Rowlands, I.H. 1995: *The politics of global atmospheric change*. Manchester: Manchester University Press.

Royal Society 1985: *The public understanding of science*. London: The Royal Society.

Rudig, W. 1995: *Public opinion and global warming*. Strathclyde papers on Government and Politics, No. 101. Glasgow: University of Strathclyde.

Santer, B.D., Wigley, T.M.L., Barnett, T.P. and Anyamba, E. 1996: Detection of climate change and attribution of causes. In Houghton, J.T., Meira Filho, L.G., Callander, B.A., Harris, N., Kattenberg, A. and Maskell, K. (eds), *Climate change 1995: the science of climate change*. Cambridge: Cambridge University Press.

SCEP 1970: *Man's impact on the global climate: report of the study of critical environmental problems*. Cambridge, MA: The MIT Press.

Schneider, S.H. 1988: The greenhouse effect and the US summer of 1988: cause and effect or media event? *Climatic Change* 13, 113–5.

Shove, E., Lutzenhiser, L., Guy, S., Hackett, B. and Wilhite, H. 1998: Energy and social systems. In Raynor, S. and Malone, E.L. (eds), *Human choice and climate change: volume 2 resources and technology*. Columbus, OH: Batelle Press.

Spencer-Cooke, A. 1999: You can't insure against the end of the world. *Green Futures* 17, July/August, 38-40.

Staats, H.J., Wit, A.P. and Midden, C.Y. 1996: Communicating the greenhouse effect to the public: evaluation of a mass media campaign from a social dilemma perspective. *Journal of Environmental Management* 46, 189–203.

Teller, E., 1984, Widespread after-effects of nuclear war. *Nature* 317, 35–9.

Walley, N. and Whitehead, B. 1994: It's not easy being green, *Harvard Business Review* May–June, 46–52.

Welford, R. and Gouldson, A. 1993: *Environmental management and business strategy*. London: Pitman Publishing.

Further reading

The following are groups of textbooks that will provide further background information. Many already appear in the reference lists.

Basic physical concepts

Bohren, C.F. 1987: *Clouds in a glass of beer: simple experiments in atmospheric physics*. New York: John Wiley and Sons.

Introductory textbooks to meteorology and climatology

Barry R.G. and Chorley, R.J. 1992: *Atmosphere, weather and climate*. 6th edition. London: Routledge.

Henderson-Sellers, A. and Robinson, P.J. 1986: *Contemporary climatology*. London: Longman Group.

Thompson, R.D. 1998: *Atmospheric processes and systems*. London: Routledge.

Meteorology and climatology

Gordon, A., Grace, W., Schwerdtfeger, P. and Byron-Scott, R. 1998: *Dynamic meteorology: a basic course*. London: Arnold.

McIlveen, R. 1992: *Fundamentals of weather and climate*. London: Chapman and Hall.

Oceanography

Duxbury, A.C. and Duxbury, A.B. 1989: *An introduction to the world's oceans.* Dubuque: Wm. C. Crown.

Thurman, H.V. 1993: *Essentials of oceanography.* 4th edition. New York: Macmillan Press.

Geology

Condie, K.C. and Sloan, R.E. 1998: *Origin and evolution of Earth: principles of historical geology.* London: Prentice-Hall.

Lowe, J.J. and Walker, M.J.C. 1997: *Reconstructing quaternary environments.* 2nd edition. Harlow: Addison Wesley Longman.

Historical climate change

Grove, J.M. 1988: *The Little Ice Age.* London: Methuen.

Lamb, H.H. 1982: *Climate history and the modern world.* London: Methuen.

Whyte, I.D. 1995: *Climatic change and human society.* London: Arnold.

Climate modelling

Trenberth, K.E. (ed.) 1992: *Climate system modeling.* Cambridge: Cambridge University Press.

McGuffie, K. and Henderson-Sellers, A. 1997: *A climate modelling primer.* 2nd edition. Chichester: John Wiley and Sons.

Introductory textbooks on global environmental change

Horel, J. and Geisler, J. 1997: *Global environmental change: an atmospheric perspective.* New York: John Wiley and Sons.

Houghton, J.T., 1994: *Global warming: the complete briefing.* Oxford: Lion Publishing.

Kemp, D.D. 1990: *Global environmental issues: a climatological approach.* London: Routledge.

Personal viewpoints

Gore, A. 1992: *Earth in the balance: forging a new common purpose*. London: Earthscan.
Lovelock, J. 1988: *The ages of Gaia*. Oxford: Oxford University Press.

Glossary

Maunder, W.J. 1994: *Dictionary of global climate change*. 2nd edition. London: UCL Press.

Glossary

absolute zero the temperature at which matter ceases to radiate. It denotes the zero point on the Kelvin temperature scale and is approximately −273.15°C.

absorption the process by which energy is removed from a radiation stream by a substance.

accretion the process by which solid particles stick together in order to form planetesimals (mini-planets).

adiabatic a process which takes place with no heat exchange between an object and its surroundings.

aerobic a process requiring oxygen (compare with **anaerobic**).

aerosols small particles of dust or compounds dissolved in water that are found in the atmosphere. Typically having a diameter of around, or less than, 1 μm. They can be anthropogenic or natural in origin.

albedo the amount of solar radiation reflected by an object compared to the total amount of incident solar radiation. It is usually measured as a fraction between 0 and 1 or as a percentage.

anaerobic without oxygen (compare with **aerobic**).

analogue models a type of model where the statistical analysis of observed data is used to provide analogues of the future. Paeleo-analogue models use reconstructed geological climates to provide analogues of global warming at future times. Historical analogue models compare series of recent past cold and warm periods to look at the difference between the two climates.

anthropogenic of human origin.

anti-cyclone an area of high atmospheric pressure.

atmosphere the gaseous envelope surrounding a planet.

atmospheric turbidity a measure of the amount of attenuation of solar radiation by the atmosphere. Dust and aerosols in the atmosphere significantly increase the atmospheric turbidity. See also **dust veil index**.

attention cycle the theory that the attention paid to an environmental issue by the public goes through distinct phases.

attribution being unequivocally able to identify the cause of a particular change in the climate. See also **detection**.

biological pump a term used to describe the process whereby carbon dioxide is transferred to the deep ocean. The processes are predominantly biological.

biosphere all living animals and plants and the matter created by them.

black body an object that absorbs all radiation incident upon it and emits all radiation over all wavelengths at a given temperature.

calving the breaking off of ice at the margins of the polar ice caps to form icebergs.

carbon dioxide fertilization a term used to describe an increase in yield observed in certain plant groups due to raised levels of atmospheric carbon dioxide.

Cartesian grid model a type of general circulation model where the **fundamental equations** are solved on a box like grid using finite difference methods.

chlorofluorocarbons (CFCs) a large group of synthetic chemicals that were principally used as coolants and aerosols propellants. They are primarily responsible for the thinning of the ozone layer and their production is prohibited under the Montreal Protocol.

climate the average state of the climate system over a period of several decades or centuries and the variability of the system about the average.

climate drift the drift away from the observed ocean climate that AOGCMs experience due to differences in the flux values calculated by the AGCM and OGCM.

climate sensitivity the scale of response of the climate system to a perturbation. Climate modellers usually refer to climate sensitivity as the change in surface air temperature in response to a radiative forcing.

climate system comprised of the five components: atmosphere, hydrosphere, cryosphere, lithosphere and biosphere. The interactions between these various components control the climate.

cold start phenomenon a term used to describe the problem that in general climate models increased radiative forcing due to increased greenhouse gases are usually applied to present-day values. However, in reality the deep oceans have been slowly responding to the build-up of gases for several hundred years.

Conference of the Parties (CoP) the name given to subsequent meetings of the signatories (parties) to the UNFCCC.

coriolis force a force generated due to the spinning of the earth. Objects moving in a straight line over the surface of the Earth are deflected right in the northern hemisphere and to the left in the southern hemisphere.

cost benefit analysis (CBA) an economic tool for assessing the overall costs and benefits of a project to society as a whole. This leads to values having to be placed on items which traditionally have no economic value. See also **willingness to pay.**

Courant-Friedrichs-Lewy (CFL) criterion as a general climate model's resolution increases, the number of time steps have to proportionally increase. This causes problems at the poles in **Cartesian grid-** based GCMs.

cryosphere glaciers, polar ice caps, seasonal snow cover; the frozen water component of the Earth–atmosphere system.

cyclone an area of low pressure, in the mid-latitudes these are often referred to as depression systems. Areas of intense low pressure can form over tropical oceans giving rise to **tropical cyclones.**

Daisyworld a computer model designed to show that in the **Gaia Hypothesis** the biosphere does not require forethought.

dendrochronolgy the determination of temperature and precipitation from the widths of tree rings.

detection showing that there has been a statistically significant change in the climate. See also **attribution.**

discount rate used in **cost benefit analysis (CBA)** to relate future costs and benefits to present-day values.

distorted physics the altered physics of the deep ocean to allow a coupled AOGCM to have a shorter **spin up time.**

double dividend see **win-win.**

downscaling the process by which large-scale **GCM** results are related to regional scale variables. This can be achieved either through using observations (empirical downscaling) or **regional climate models.**

dust veil index (DVI) an index to assess the affect of volcanic eruptions on the amount of solar radiation passing through the atmosphere. Krakatoa in 1893 is assigned a DVI of 1000.

eccentricity a measure of the deviation of an ellipse from a circle. The eccentricity of the Earth's orbit varies over time.

ecological modernization theory (EMT) a theory that promotes the idea that the challenges that achieving sustainable development presents are best solved by the application of technology and science.

Ekman spiral an effect due to friction which occurs in both the ocean and atmosphere. The surface layer of the ocean is dragged along by surface winds. Lower layers of the ocean are less affected and the direction of flow spirals round. In the atmosphere the surface provides a frictional drag. The free atmospheric flow is increasingly affected by this drag the nearer it is to the surface. Again, this creates a spiral in the wind flow.

El Niño this usually refers to the oceanic changes which accompany the **ENSO**.

El Niño Southern Oscillation (ENSO) this describes an irregular oscillation of the climate system in the Pacific Ocean. It is linked to anomalously high sea surface temperatures off the coast of Peru (**El Niño**) coupled with a change in the atmospheric pressure regime (**Southern Oscillation**).

emissivity a measure of how well an object radiates energy compared to a black body at the same temperature.

equilibrium the balance point between the various forces acting on the climate system.

Equatorial Trough band of low pressure associated with the **Intertropical Convergence Zone (ITCZ)**.

evaluation: the term used from the 1995 IPCC report to describe the process of checking a GCM's ability to model the present-day climate. See also **validation**.

extinction the combined processes of absorption and scattering.

faint sun paradox this describes the paradoxical geological evidence that liquid water existed on Earth at a time when the solar luminosity was so low that the Earth should have been frozen. The paradox is usually solved by invoking an enriched carbon dioxide atmosphere so leading to an enhanced greenhouse effect.

feedback within a system a change in the signal results in changes in the output which then feeds back into the signal again. Positive feedbacks amplify the original change whereas negative feedbacks ameliorate any change.

flux adjustment an alteration to the fluxes between a coupled AGCM and OGCM to correct **climate drift**.

forcing a change in the net radiation balance.

fossil fuels these are fuels that had a biological origin.

fundamental equations the basic laws of physics that govern the motion of fluids which are used in general circulation models to model the atmosphere and oceans. It also includes the conservation of moisture and other constituents (compare with **primitive equations**).

Gaia hypothesis a theory promoting the view that the Earth's climate is regulated by the biosphere to enhance the survival of life.

general circulation model (GCM) a computer-based model which attempts to predict the general circulation of the atmosphere (AGCM) or the ocean (OGCM).

general climate model (GCM) computer-based models that attempt to simulate the climate from first principles using a suite of coupled models. Note the abbreviation GCM was widely used to mean **general circulation model** prior to the early 1990s.

global warming a term used to describe the change in annual average global surface temperature, and related climatic changes, due to increases in greenhouse gases, most notably carbon dioxide, from anthropogenic emissions.

global warming potential (GWP) the cumulative radiative forcing between the present time and some later time horizon caused by a unit of gas emitted now, considered relative to some reference gas. Usually the reference gas used is carbon dioxide.

greenhouse effect the warming at the Earth's surface due to atmospheric greenhouse gases absorbing outgoing long-wave radiation. This radiation is then re-emitted and thereby warms the Earth's surface.

greenhouse gases atmospheric gases that absorb outgoing long-wave radiation. In the Earth's atmosphere the two most important greenhouse gases are water vapour and carbon dioxide.

growing season when the growth of plants is not restricted by temperature.

Hadley cell a convective cell driven by the thermal heat of the sun. One exists on either side of the equator. The air rises at the equator, travels polewards and sinks around 30° latitude to travel equatorward at the surface.

half-life the time taken for one half of the original number of radioactive atoms to decay.

halocline the layer of the ocean, from around 100m to 1000m, where the salinity changes rapidly with depth.

hydrosphere the oceans, lakes and rivers; all the liquid water part of the Earth–atmosphere system. Note that clouds are usually included as part of the atmosphere.

ice age a rather vague term with no strict definition. It is most commonly associated with quaternary glacial cycles and the encroachment of glacial ice sheets at lower latitudes.

ideal gas a gas where the molecules are far enough apart so as not to interact with each other. Although an ideal gas is essentially hypothetical the atmosphere at room temperature closely approximates one.

impact assessment models (IAM) see **integrated assessment models.**

insolation the total amount of incoming solar radiation over a unit area at a particular time and place at the Earth's surface.

integrated assessment model (IAM) also called impact assessment models. These models attempt to encompass a wide range of global warming impacts and include a suite of models which encompass climate change and economic impacts. They have been put to a wide range of uses.

interglacial describes periods of comparative warmth during the quaternary when the ice sheet has shrunk back to polar regions.

Intergovernmental Panel on Climate Change (IPCC) an international organization established by WMO and UNEP. There are three working groups made up of individuals nominated by member governments. Working group 1 reviews the scientific evidence related to global warming. Working group 2 reviews the social and economic impacts of global warming. Working group 3 reviews mitigation and reduction strategies.

Intertropical Convergence Zone (ITCZ) a band of preferred convergence that occurs around the equator. It is marked by a band of clouds, showing rising air, but does not completely encircle the globe.

isotope an element having the same number of protons but a different number of neutrons.

jet stream a ribbon-like structure of fast moving air in the upper troposphere.

Kelvin wave the equatorial Kelvin wave associated with the **ENSO** is caused by changes in the surface winds. It occurs in the equatorial upper ocean and travels eastwards (compare with **Rossby waves**).

kinetic energy the energy an object has due to its movement.

La Niña a time of anomalously cold sea surface temperatures off the South American coast (compare with **El Niño**).

lapse rate the rate of change of a variable with altitude, usually that variable is temperature.

latent heat the heat required to change 1g mass of a substance from one state to another without a change in temperature.

leaching a process whereby water moving downwards through a soil dissolves nutrients and pollutants contained in the soil. The dissolved solutes are then washed away or moved to lower soil layers.

leads long but narrow breaks in sea ice.

limited area models (LAMs) see **regional climate models.**

linear equation an equation that describes a straight line. The dependent variable will be to the power of 1.

lithosphere the solid part of the Earth's surface.

Little Ice Age a term used to describe a period of cooling in the middle of the second millennium. It is mostly associated with low temperatures over Europe during the 1500s and 1600s.

Maunder minimum a period of about 50 years in the second half of the 1600s when the Sun appeared not to have any sunspots. It coincides with the **Little Ice Age.** Other periods when the Sun is believed to have been free of sunspots have also occurred.

meridional denoting longitude bands, i.e. running north–south.

mesosphere the upper portion of the middle atmosphere extending from around 50 km to 80 km. It is topped by the mesopause.

Milankovitch cycles a name given to the three variations in the Earth's orbit: **eccentricity, obliquity** and **precession.**

monsoon a term used to describe the seasonal change in wind direction over the Indian sub-continent. It is now used to describe such wind patterns in other parts of the globe such as Africa and America.

Montreal Protocol an international treaty signed in 1987 which limited the production of ozone depleting chemicals. This treaty has been continuously updated.

noise natural variations in the climate which may mask the climate change being looked for, such as global warming.

non-governmental organization (NGO) a term used to describe a wide variety of organizations that have no government involvement. These include such organizations such as Greenpeace and Friends of the Earth.

non-linear equation an equation containing terms of the dependent variable other than to the power of one. In other words, the variables do not increase linearly.

no regrets policy the name given to the policy of adopting measures that as well as tackling global warming also have other social benefits. Therefore even if global warming does not emerge, benefits to society will accrue anyway.

North Atlantic Deep Water (NADW) an important cold flow of the **oceanic conveyor belt** which forms south of Greenland.

North Atlantic Oscillation (NAO) a change in atmospheric pressure over the North Atlantic Ocean which bears similarities to the **ENSO**. It causes changes in the weather patterns over Europe. The North Atlantic Oscillation Index measures the pressure difference between Iceland and the Azores.

north–south terminology that is now widely used in preference to First and Third Worlds. North is taken to mean the developed countries and South the developing countries.

nuclear winter a term used to described the global cooling predicted to occur after a major nuclear war

obliquity the tilt of the Earth's axis.

oceanic conveyor belt a term used to describe the circulation of the deep ocean. The upper part of this circulation consists of warm surface water and the lower part is cold bottom water. In the present day it acts as a pump transferring heat from the low latitudes to the high latitudes.

outgassing the release of volatiles from rocks which would have formed the secondary planetary atmospheres of the terrestrial planets.

ozone depletion potential (ODP) the amount of ozone destroyed by the emission of a gas over its entire lifetime in the atmosphere compared to that destroyed by the emission of the same mass of CFC-11.

ozone hole a term used to describe a thinning of the ozone layer over Antarctica. A similar thinning, though not as deep, has been discovered over the Arctic.

ozone layer a name given to the peak concentration of ozone in the atmosphere. It occurs in the stratosphere between 15 and 40 km altitude.

parameterization the empirically derived equations used in general climate models to represent sub-grid scale processes.

pH a measure of how acid or alkaline a solution is. The scale ranges from 1 (most acidic such as hydrochloric acid) to 14 (most alkaline such as sodium hydroxide). A pH value of 7 is considered neutral, neither acid nor alkaline. It is a logarithmic scale so a pH value of 2 is ten times more acidic than a pH value of 3.

photodissociation the breaking up of a molecule into smaller constituents due to the action of sunlight.

photosynthesis the conversion by plants of carbon dioxide and water into organic compounds using sunlight.

polynyas large areas of open water in sea-ice.

potential energy energy that is stored by an object. It can take various forms such as gravitational and chemical.

precautionary principle a new process or material should not be developed unless it can be proved that it does not harm the environment. Also adequate financial provision is made to rectify any unseen future harm.

precession the movement of the equinoxes.

primitive equations the **fundamental equations** assuming hydrostatic equilibrium. That is they do not include equations for the conservation of water or other constituents.

proxy data data which bear some relationship to the climate and therefore can be interpreted to reveal climate data.

pycnocline the layer of the ocean, from around 100 m to 1000 m, where the salinity and temperature changes rapidly with depth (**halocline** and **thermocline**).

radiative forcing a change in the balance between the incoming and outgoing radiation, that is the average net radiation, at the tropopause.

radical an atom, or group of atoms containing at least one unpaired electron.

radiocarbon dating the dating of fossils or other preserved minerals by looking at radioisotopes of carbon.

regional climate models (RCMs) a high resolution model concentrated on regional variables. They are forced at their boundaries by the results of GCMs. They are also called **limited area models (LAMs)**.

relaxation time see **response time**.

response time the time taken for a component of the climate system to return to equilibrium after a change in the forcing conditions. Also called the **relaxation time**.

risk society theory a complex philosophical theory that considers many aspects of society as it leaves the industrial (modern) era. Modern society is seen as being faced by catastrophic risks caused primarily by science and technology. The individual is perceived as powerless against experts and unable to escape the risks.

Rossby waves usually used to describe fast-flowing undulating westerlies in the upper troposphere. The main Rossby flow occurs between 35° and 55° latitude. They arise because of vorticity and the coriolis force. Equatorial Rossby waves are associated with the **ENSO** and occur in the equatorial upper ocean and travel westwards (compare with **Kelvin wave**).

salinity a measure of the dissolved salts in the ocean.

scattering a process in which the direction and intensity of radiation is altered by a substance.

sensitivity a term used to relate the surface temperature change in response to a change in radiative forcing. It takes account of all the feedback processes within the climate system. The sensitivity of the climate system to a radiative forcing, such as a change in greenhouse gases, is pivotal to the question of global warming. Sceptics claim that climate models over-estimate the sensitivity of the climate system.

solar constant the amount of solar energy incident upon a unit surface area per unit time at the top of the atmosphere.

solar wind a stream of charged particles that constantly leaves the Sun. It is believed that it was much stronger in the early history of the solar system.

South Pacific Convergence Zone (SPCZ) another noted area of convergence (see **Intertropical Convergence Zone**), this time in the south Pacific extending from Indonesia south-eastwards into the central south Pacific.

Southern Oscillation changes in atmospheric pressure over the Pacific basin associated with the El Niño. The southern oscillation index measures the pressure between the Indonesian lows and Pacific highs.

specific heat capacity the heat required to raise the temperature of 1g mass of a substance by 1°C.

spectral method a type of general circulation model where the fundamental equations are solved as waveforms in the horizontal plane. Compare to **Cartesian grid model**.

spin-up time the time taken for a climate model to reach realistic climate conditions.

Stefan-Boltzmann law the relationship between the rate at which energy is emitted and temperature.

stratosphere the lowest part of the middle atmosphere. It extends from approximately 12 km to 45 km and contains the **ozone layer**.

sunk cost a term applied in **cost benefit analysis** to describe costs that were spent in the past; these are seen as unchangeable.

sunspot number the number of sunspots multiplied by ten added to the number of individual sunspots multiplied by a constant.

sustainable development a rather unclear concept although often it is interpreted from the following working definition. That sustainable development should meet the needs of the present without compromising the ability of future generations to meet their own needs.

teleconnections a term used to describe how changes in weather patterns in one part of the globe can be propagated through the atmosphere to affect the weather in other regions of the world.

temperature the mean kinetic energy per molecule of all the molecules in an object.

time step defines the temporal resolution of a **GCM**.

thermocline the layer of the ocean, from around 100m to 1000m, where the temperature decreases rapidly with depth.

trade winds extremely constant winds that blow back to the equator from the sub-tropics. They are deflected eastwards due to the **coriolis force**.

transient used to describe the climate system which is still responding to a forcing. A transient climate will represent part of the **equilibrium** response.

tropical cyclones the generic term for intense storms that affect tropical areas. They are known in different regions under local names. In the Atlantic and eastern North Pacific they are called hurricanes. In the western North Pacific they are called Typhoons.

troposphere the lowest part of the atmosphere. Extends to around 10 to 15 km in altitude depending on latitude and season. It is topped by the tropopause.

turbidity a measure of the amount of attenuation of solar radiation by the atmosphere.

urban heat island a term used to describe the effect of urban areas on the natural climate. Typically a rise in temperature is seen in built-up areas compared to the surrounding countryside. Temperature is just one of many climatic variables to be affected by the urban canopy.

United Nations Framework on Climate Change (UNFCC) signatories to this document agreed to stabilize greenhouse gases in the atmosphere at a level that would not endanger the climate system.

validation the term used prior to the 1995 IPCC report to describe the process of checking a GCM's ability to model the present-day climate. This term has been superseded by the phrase **evaluation**.

Van Allen radiation belts these are composed of highly energetic, charged particles from cosmic rays which become trapped in the Earth's magnetic field.

varve layers of sediment laid down by lakes during a year. Each layer is composed of two segments, a light and a dark layer formed during spring–summer and autumn–winter respectively.

wait and see the policy of pursuing more research to reduce the scientific uncertainty associated with global warming rather than implementing policies to curb greenhouse gas emissions now.

weather the state of the atmosphere at a particular location and time.

willingness to pay (WTP) an economic tool for placing a value on an object which does not traditionally have a monetary value. It is the amount people are willing to pay to preserve an ecosystem or animal.

win-win traditionally business and care for the environment is seen as incompatible. The win-win philosophy argues that extra dividends accrue to business that take care of the environment. It is also referred to as the **double dividend**.

Younger Dryas a period of pronounced cooling that occurred as the world recovered from the last quaternary ice age. It probably lasted for around 700 years and ended very abruptly.

zonal denoting latitude bands, i.e. running east–west.

Useful websites

These websites have been very loosely grouped to indicate the main function of the listed institution.

Scientific institutes

IPCC
http://www.ipcc.ch/about/about.htm
Contains information about the IPCC, the various working groups with links to press releases, publications etc.

WMO
http://www.wmo.ch/
World Meteorological Organisation

International Institute for Applied Systems Analysis (IIASA)
http://www.iiasa.ac.at/
A privately funded (from charities and UNEP) institute particularly linking east and west research initiatives.

National Climatic Data Center
http://www.ncdc.noaa.gov/#ABOUT
Provides climate data.

UK Meteorological Office
http://www.meto.govt.uk/
The government funded weather service for the UK it also has a branch dedicated to climate change studies called the Hadley Centre whose web site address is
http://www.meto.govt.uk/sec5/CR_div/Brochure/

Inter-governmental agencies

UNEP
http://www.unep.ch/
This site provides links to the UNFCCC and IPCC.

European environment agency
http://www.eea.eu.int

Non-governmental organizations

Pro-emission reduction

Greenpeace International: Climate
http://www.greenpeace.org/~climate
Contains rebuttals of the main anti-global warming arguments.

Friends of the Earth
http://www.foe.co.uk/climatechange/
Contains links to other sites regarding climate change including those against
global warming initiatives.

Anti-emission reduction

Institute of Economic Affairs
http://www.iea.org.uk/
A UK based think-tank promoting free market trade.

Global Climate Coalition
http://www.globalclimate.org/
Perhaps the main US voice for businesses and industry lobbying against
global warming initiatives. Contains links to other like-minded organizations.

George C. Marshall Institute
http://www.marshall.org/
US-based institute.

CATO Institute
http://www.cato.org/
A US organization promoting deregulation of industry, limited government
and free markets.

validation. *See* evaluation
Van Allen radiation belts, 116
varves, 84, 106
Venus: early atmosphere, 73
 life, 75
 runaway greenhouse effect, 74
Vikings, 127
volcanic activity, 120
 geological, 90
vorticity, 37

wait-and-*see* approach, 213, 222
Walker Circulation, 34
water: structure of water molecule, 19
water vapour: absorption bands, 21,
 150

positive feedback effect, 195
wavelength, 10–22, 150
weather, 1, 5, 29, 31, 33, 105, 127
 extreme events, 158
 numerical weather prediction, 172
West Antarctic Ice Sheet, 158
willingness to pay (WTP), 209, 213
wind, 44
win-win. *See* double dividend
Wolf Minimum, 116
Wolf-Gleissberg Cycle, 118
work, 8
World Meteorological Organization
 (WMO), 2

Younger Dryas, 97